IAEA NUCLEAR SAFETY
AND SECURITY GLOSSARY

The following States are Members of the International Atomic Energy Agency:

AFGHANISTAN
ALBANIA
ALGERIA
ANGOLA
ANTIGUA AND BARBUDA
ARGENTINA
ARMENIA
AUSTRALIA
AUSTRIA
AZERBAIJAN
BAHAMAS
BAHRAIN
BANGLADESH
BARBADOS
BELARUS
BELGIUM
BELIZE
BENIN
BOLIVIA, PLURINATIONAL
    STATE OF
BOSNIA AND HERZEGOVINA
BOTSWANA
BRAZIL
BRUNEI DARUSSALAM
BULGARIA
BURKINA FASO
BURUNDI
CAMBODIA
CAMEROON
CANADA
CENTRAL AFRICAN
    REPUBLIC
CHAD
CHILE
CHINA
COLOMBIA
COMOROS
CONGO
COSTA RICA
CÔTE D'IVOIRE
CROATIA
CUBA
CYPRUS
CZECH REPUBLIC
DEMOCRATIC REPUBLIC
    OF THE CONGO
DENMARK
DJIBOUTI
DOMINICA
DOMINICAN REPUBLIC
ECUADOR
EGYPT
EL SALVADOR
ERITREA
ESTONIA
ESWATINI
ETHIOPIA
FIJI
FINLAND
FRANCE
GABON
GEORGIA

GERMANY
GHANA
GREECE
GRENADA
GUATEMALA
GUYANA
HAITI
HOLY SEE
HONDURAS
HUNGARY
ICELAND
INDIA
INDONESIA
IRAN, ISLAMIC REPUBLIC OF
IRAQ
IRELAND
ISRAEL
ITALY
JAMAICA
JAPAN
JORDAN
KAZAKHSTAN
KENYA
KOREA, REPUBLIC OF
KUWAIT
KYRGYZSTAN
LAO PEOPLE'S DEMOCRATIC
    REPUBLIC
LATVIA
LEBANON
LESOTHO
LIBERIA
LIBYA
LIECHTENSTEIN
LITHUANIA
LUXEMBOURG
MADAGASCAR
MALAWI
MALAYSIA
MALI
MALTA
MARSHALL ISLANDS
MAURITANIA
MAURITIUS
MEXICO
MONACO
MONGOLIA
MONTENEGRO
MOROCCO
MOZAMBIQUE
MYANMAR
NAMIBIA
NEPAL
NETHERLANDS
NEW ZEALAND
NICARAGUA
NIGER
NIGERIA
NORTH MACEDONIA
NORWAY
OMAN
PAKISTAN

PALAU
PANAMA
PAPUA NEW GUINEA
PARAGUAY
PERU
PHILIPPINES
POLAND
PORTUGAL
QATAR
REPUBLIC OF MOLDOVA
ROMANIA
RUSSIAN FEDERATION
RWANDA
SAINT KITTS AND NEVIS
SAINT LUCIA
SAINT VINCENT AND
    THE GRENADINES
SAMOA
SAN MARINO
SAUDI ARABIA
SENEGAL
SERBIA
SEYCHELLES
SIERRA LEONE
SINGAPORE
SLOVAKIA
SLOVENIA
SOUTH AFRICA
SPAIN
SRI LANKA
SUDAN
SWEDEN
SWITZERLAND
SYRIAN ARAB REPUBLIC
TAJIKISTAN
THAILAND
TOGO
TONGA
TRINIDAD AND TOBAGO
TUNISIA
TÜRKİYE
TURKMENISTAN
UGANDA
UKRAINE
UNITED ARAB EMIRATES
UNITED KINGDOM OF
    GREAT BRITAIN AND
    NORTHERN IRELAND
UNITED REPUBLIC
    OF TANZANIA
UNITED STATES OF AMERICA
URUGUAY
UZBEKISTAN
VANUATU
VENEZUELA, BOLIVARIAN
    REPUBLIC OF
VIET NAM
YEMEN
ZAMBIA
ZIMBABWE

The Agency's Statute was approved on 23 October 1956 by the Conference on the Statute of the IAEA held at United Nations Headquarters, New York; it entered into force on 29 July 1957. The Headquarters of the Agency are situated in Vienna. Its principal objective is "to accelerate and enlarge the contribution of atomic energy to peace, health and prosperity throughout the world".

IAEA NUCLEAR SAFETY AND SECURITY GLOSSARY

TERMINOLOGY USED IN NUCLEAR SAFETY, NUCLEAR SECURITY, RADIATION PROTECTION AND EMERGENCY PREPAREDNESS AND RESPONSE

2022 (INTERIM) EDITION

INTERNATIONAL ATOMIC ENERGY AGENCY
VIENNA, 2022

COPYRIGHT NOTICE

For further information on this publication, please contact:

Safety Standards and Security Guidance Development Section
International Atomic Energy Agency
Vienna International Centre
PO Box 100
1400 Vienna, Austria
Email: Official.Mail@iaea.org

© IAEA, 2022
Printed by the IAEA in Austria
October 2022

IAEA Library Cataloguing in Publication Data

Names: International Atomic Energy Agency.
Title: IAEA nuclear safety and security glossary / International Atomic Energy
Agency.
Description: Vienna : International Atomic Energy Agency, 2022. | Includes bibliographical
references.
Identifiers: IAEAL 22-01539 | ISBN 978–92–0–141822–7 (paperback : alk. paper) |
ISBN 978–92–0–141122–8 (pdf) |
Subjects: LCSH: | Nuclear industry — Safety regulations. | Nuclear industry — Security
measures. | Nuclear industry — Glossaries, vocabularies, etc.
Classification: UDC 621.039.58 (038) | IAEA/NSS/GLO

FOREWORD

In developing and establishing standards of safety and guidance on nuclear security, clear communication on scientific and technical terms and concepts is essential. These terms and concepts are established and defined in individual IAEA safety standards and nuclear security guidance. It is helpful to further explain to readers these terms and their usage. To this end, the IAEA Nuclear Safety and Security Glossary reproduces these definitions and explains their usage in IAEA safety standards, nuclear security guidance and other safety and security related IAEA publications.

The 2022 (Interim) Edition of the IAEA Nuclear Safety and Security Glossary presents a collection of terms and definitions that have been established in IAEA safety standards and nuclear security guidance issued and approved up to 2022. The primary purpose of the IAEA Nuclear Safety and Security Glossary is to promote the consistent usage of terms in IAEA safety standards and nuclear security guidance. Once definitions of terms have been established, they are intended to be observed in IAEA safety standards and nuclear security guidance and other safety and security related IAEA publications.

The IAEA Nuclear Safety and Security Glossary provides guidance primarily for drafters and reviewers of safety standards, including IAEA technical officers and consultants, and bodies for the endorsement of safety standards. It is also a source of information for users of IAEA safety standards and nuclear security guidance and other safety and security related IAEA publications and for other IAEA staff — notably writers, editors, translators, revisers and interpreters. Users of the IAEA Nuclear Safety and Security Glossary, in particular drafters of national legislation, should be aware that terminology and usage may differ in other contexts, such as in binding international legal instruments and in publications from other organizations.

It is recognized that there are several terms and definitions included in this 2022 (Interim) Edition of the IAEA Nuclear Safety and Security Glossary, particularly in the area of nuclear security where differences in usage still need to be reconciled and further harmonization is needed. Significant efforts are currently under way to resolve these terminological issues. The IAEA intends to revise and update this glossary when these efforts are complete or when new or revised Safety Requirements and Nuclear Security Recommendations publications containing new terms and definitions are issued.

The IAEA invites users of this glossary or of other safety and security related IAEA publications to submit comments and suggestions on the definitions of terms and their explanations for consideration in a future edition of the IAEA Nuclear Safety and Security Glossary.

The IAEA officer responsible for this publication was K. Asfaw of the Office of Safety and Security Coordination.

CONTENTS

INTRODUCTION

BACKGROUND

Terminology in IAEA safety standards

The IAEA safety standards for nuclear installations, radiation protection and safety, radioactive waste management and the transport of radioactive material were historically developed in four separate programmes. For nuclear installations and radioactive waste management, safety standards programmes were set up to coordinate the development of standards covering the different areas of each subject. The radiation and transport safety standards programmes were each centred on one key set of safety requirements — the Basic Safety Standards (the current edition of which is IAEA Safety Standards Series No. GSR Part 3, Radiation Protection and Safety of Radiation Sources: International Basic Safety Standards [1]) and the Transport Regulations (the current edition of which is IAEA Safety Standards Series No. SSR-6 (Rev. 1), Regulations for the Safe Transport of Radioactive Material, 2018 Edition [2]), respectively — with other safety requirements and guidance elaborating on particular parts of these central publications. At the outset, each of the four groups of safety standards had developed its own terminology:

(a) In nuclear safety, compilations of terms and definitions were produced for internal use but not published. Nevertheless, the lists of definitions given in the Nuclear Safety Standards Codes published by the IAEA in 1988[1] provided a set of the fundamental terms.

(b) In 1986, the IAEA published a Radiation Protection Glossary[2], which provided a collection of fundamental terms associated with radiation protection, along with their definitions. Many of the terms and definitions in this publication are now obsolete, and the Basic Safety Standards issued in 1996[3] (superseded in 2014 [1]) included more up to date definitions of terms in radiation protection and safety.

(c) In 1982, a Radioactive Waste Management Glossary was published by the IAEA as IAEA-TECDOC-264[4]. A revised and updated version was issued in 1988 as IAEA-TECDOC-447[5], a third edition was published in 1993[6] and a fourth edition was published in 2003 [3].

---

[1] INTERNATIONAL ATOMIC ENERGY AGENCY, Code on the Safety of Nuclear Power Plants: Governmental Organization, Safety Series No. 50-C-G (Rev. 1), IAEA, Vienna (1988).
INTERNATIONAL ATOMIC ENERGY AGENCY, Code on the Safety of Nuclear Power Plants: Siting, Safety Series No. 50-C-S (Rev. 1), IAEA, Vienna (1988).
INTERNATIONAL ATOMIC ENERGY AGENCY, Code on the Safety of Nuclear Power Plants: Design, Safety Series No. 50-C-D (Rev. 1), IAEA, Vienna (1988).
INTERNATIONAL ATOMIC ENERGY AGENCY, Code on the Safety of Nuclear Power Plants: Operation, Safety Series No. 50-C-O (Rev. 1), IAEA, Vienna (1988).
INTERNATIONAL ATOMIC ENERGY AGENCY, Code on the Safety of Nuclear Power Plants: Quality Assurance, Safety Series No. 50-C-QA (Rev. 1), IAEA, Vienna (1988).
[2] INTERNATIONAL ATOMIC ENERGY AGENCY, Radiation Protection Glossary (Safety Guide), Safety Series No. 76, IAEA, Vienna (1986).
[3] FOOD AND AGRICULTURE ORGANIZATION OF THE UNITED NATIONS, INTERNATIONAL ATOMIC ENERGY AGENCY, INTERNATIONAL LABOUR ORGANISATION, PAN AMERICAN HEALTH ORGANIZATION, OECD NUCLEAR ENERGY AGENCY, WORLD HEALTH ORGANIZATION, International Basic Safety Standards for Protection against Ionizing Radiation and for the Safety of Radiation Sources, Safety Series No. 115, IAEA, Vienna (1996).
[4] INTERNATIONAL ATOMIC ENERGY AGENCY, Radioactive Waste Management Glossary, IAEA-TECDOC-264, IAEA, Vienna (1982).
[5] INTERNATIONAL ATOMIC ENERGY AGENCY, Radioactive Waste Management Glossary, Second Edition, IAEA-TECDOC-447, IAEA, Vienna (1988).
[6] INTERNATIONAL ATOMIC ENERGY AGENCY, Radioactive Waste Management Glossary, IAEA, Vienna (1993).

(d) The definitions in the IAEA Transport Regulations, 2018 Edition [2] represent current terminology for transport safety.

With the creation of the Department of Nuclear Safety in 1996, and the adoption of a harmonized procedure for the preparation and review of safety standards in all areas, the need for greater consistency in the use of terminology became apparent. The IAEA Safety Glossary was first published in 2007[7] and was intended to contribute towards harmonizing the use of terminology in IAEA safety standards and the IAEA's other safety and security related publications. A revised edition of the IAEA Safety Glossary was issued in 2018[8], to take into account new terminology and usage in Safety Requirements publications issued between 2007 and 2018. Although no Safety Requirements publications have been issued since 2018, the safety related terminology included in this 2022 (Interim) Edition of the IAEA Nuclear Safety and Security Glossary has been expanded to include terms and definitions established and documented in all currently valid Safety Guides. This 2022 (Interim) Edition of the IAEA Nuclear Safety and Security Glossary supersedes the 2018 Edition of the IAEA Safety Glossary.

Terminology in IAEA nuclear security guidance

The IAEA's nuclear security guidance began with the publication of INFCIRC/225 in 1975, providing guidance for States on the physical protection of nuclear material, which was further revised and reissued several times through the 1970s, 1980s and 1990s. INFCIRC/225 came to be used by some States Parties to the Convention on the Physical Protection of Nuclear Material (CPPNM) [4, 5] as guidance to assist them in meeting their obligations under the Convention, and the terminology used in INFCIRC/225 was largely the same as that in the Convention. Some guidance was also developed in the late 1990s relating to the security of radioactive sources, but largely as an extension to guidance on the safety of such sources and using some of the terminology of radiation protection. Since the adoption of the IAEA's first Nuclear Security Plan in 2002, the scope of nuclear security has been broadened to cover other aspects of the security of nuclear material and nuclear facilities, such as material accounting and control and computer security; the security of other radioactive material and associated facilities and activities; and security for nuclear and other radioactive material out of regulatory control.

The first publication in the IAEA Nuclear Security Series was issued in 2006. Further Implementing Guides and Technical Guidance publications on specific technical topics were issued in the following years (some of which have recently been updated). Nuclear Security Recommendations for nuclear material and nuclear facilities [6], for radioactive material and associated facilities [7] and for nuclear and other radioactive material out of regulatory control [8] were issued in 2011, followed by further specific Implementing Guides and Technical Guidance. Nuclear Security Fundamentals were published in 2013 [9], and further Implementing Guides and Technical Guidance on a range of topics have been published in recent years. The terminology used in the IAEA Nuclear Security Series has been documented progressively in individual guidance publications, through lists of definitions, footnotes and descriptions in the text. As the set of Nuclear Security Fundamentals, Recommendations and Implementing Guides (and therefore the main set of terminology for the guidance) is now largely complete, this 2022 (Interim) Edition of the IAEA Nuclear Safety and Security Glossary contains the terminology of this first iteration of the full Nuclear Security Series.

---

[7] INTERNATIONAL ATOMIC ENERGY AGENCY, IAEA Safety Glossary: Terminology Used in Nuclear Safety and Radiation Protection, 2007 Edition, IAEA, Vienna (2007).

[8] INTERNATIONAL ATOMIC ENERGY AGENCY, IAEA Safety Glossary: Terminology Used in Nuclear Safety and Radiation Protection, 2018 Edition, IAEA, Vienna (2019).

OBJECTIVE

The IAEA Nuclear Safety and Security Glossary does not itself define the terms used in safety and nuclear security. Instead, it represents a comprehensive collection of terms already defined in published safety standards and nuclear security guidance. As such, it serves a number of different purposes:

(a) To explain the meanings of technical terms that may be unfamiliar to the reader;

(b) To explain any special meanings ascribed to common words or terms (since words can have several different meanings, it may be necessary to clarify which meaning is intended, in particular for non-native English speakers);

(c) To define precisely how specific terms — whose general meaning may be clear to readers — are used in a particular context, in order to avoid ambiguity concerning some important aspect(s) of their meaning;

(d) To explain the connections or differences between similar or related terms, or the specific meanings of the same technical term in different contexts;

(e) To clarify and, if possible, reconcile differences in the usage of specialized terms in different subject areas, since such differences in usage may be potentially misleading;

(f) To recommend terms that should be used in IAEA publications and documents (and identify those that should not), and provide the definitions that should be ascribed to them, and thereby promote consistent usage in IAEA publications;

(g) To facilitate the translation of IAEA publications.

Definitions of the type used in legal texts such as the Convention on Nuclear Safety [10], the Joint Convention on the Safety of Spent Fuel Management and on the Safety of Radioactive Waste Management [11], the CPPNM and its 2005 Amendment [4, 5], and the International Convention for the Suppression of Acts of Nuclear Terrorism (ICSANT) [12], and in regulations such as the Transport Regulations [2], are intended primarily for purpose (c) and, in some cases, do not serve the other purposes at all. Furthermore, definitions of this nature tend to be tailored to the needs of the specific text to which they relate, and hence are often not generally applicable.

A glossary is not the place to specify requirements or guidance. The definition of a term contains the conditions to be met in order for the term to be applicable, but not other conditions. This is best illustrated by an example. The definition of *regulatory body* indicates the conditions that need to be met in order for an organization to be described as a *regulatory body*, but not the attributes of a *regulatory body* as required by IAEA safety standards. Hence, the definition specifies that it is "designated by the government of a State as having legal authority for conducting the regulatory *process*" — otherwise, it is not a *regulatory body*. However, the definition does not specify that it is "**independent in its safety related decision making and that it has functional separation from entities having responsibilities or interests that could unduly influence its decision making**" [13] — it can be a *regulatory body* without being independent, even though it would then not satisfy the IAEA safety requirements on legal and governmental infrastructure for safety.

SCOPE

The IAEA Nuclear Safety and Security Glossary is necessarily limited in its scope, as it is intended to focus on the key terms that are specific to, or that are used in a specific way in, protection and safety and nuclear security. A number of general categories of terms that may be used in nuclear safety and security related IAEA publications have been specifically excluded from the IAEA Nuclear Safety and Security Glossary (except where a specific point needs to be made about a specific term). These groups of excluded terms include:

(a) Basic terms from radiation and nuclear physics (e.g. alpha particle, decay, fission, radionuclide). An understanding of these terms is assumed.

(b) Terminology from safeguards that is addressed in the IAEA Safeguards Glossary [14]. Such terms and definitions may in some cases be referred to or discussed in the IAEA Nuclear Safety and Security Glossary, but the IAEA Safeguards Glossary should be consulted where it is the appropriate authority.

(c) The specialized terminology of fields other than protection and safety and nuclear security (e.g. geology, seismology, meteorology, medicine, criminology, intelligence, computing). This terminology may be used in the contexts of protection and safety and nuclear security, but the definition of such terms is left to the experts in the relevant fields.

(d) Detailed, very specialized terminology from a specific field within protection and safety and nuclear security (e.g. the detailed terminology of dosimetry and performance testing of equipment). If necessary, such terminology is defined in the specialized publications to which it is relevant. For this reason, terms and definitions listed in Technical Guidance publications are not reproduced in the IAEA Nuclear Safety and Security Glossary.

In the context of the IAEA's Major Programme on Nuclear Safety and Security, 'protection and safety' denotes the protection of people and the environment against radiation risks, and the safety of facilities and activities that give rise to radiation risks. 'Protection and safety' encompasses the safety of nuclear installations, radiation safety, the safety of radioactive waste management and safety in the transport of radioactive material; it does not include aspects of safety that are not related to radiation protection and nuclear safety. 'Nuclear safety' is usually abbreviated to 'safety' in IAEA publications on nuclear safety, and 'safety' means 'nuclear safety' unless otherwise stated. Safety is concerned with both radiation risks under normal circumstances and radiation risks as a consequence of incidents.

'Nuclear security' denotes the prevention and detection of, and response to, theft, sabotage, unauthorized access, illegal transfer or other malicious acts involving nuclear material, other radioactive substances or their associated facilities. 'Nuclear security' is often abbreviated to 'security' in IAEA publications on nuclear security.

Safety measures and security measures have in common the aim of protecting human life and health and the environment. Safety measures and security measures have to be designed and implemented in an integrated manner to develop synergy between these two areas, and also in such a way that security measures do not compromise safety and safety measures do not compromise security.

Although the ultimate aim is the same, and some features are common to both, the general approaches and specific measures taken for safety and security are often different. Furthermore, safety standards and nuclear security guidance have until recently been developed separately from different bases, and have been published in different series. Therefore, there is potential for confusion in the use of certain terminology, particularly with regard to terms used in publications that address related aspects of both safety and security:

(a) Terms that are used in both safety standards and nuclear security guidance with essentially the same meaning, but that have different definitions. For example, in both safety and security *defence in depth* is an approach whereby multiple layers of measures are applied to reduce the likelihood of undesirable consequences on the basis that, if one layer fails for any reason, other layers can still prevent the undesirable consequences. The two formal definitions reflect the two different sets of undesirable consequences with which each area is respectively concerned: the evolution of an accident or the degradation of nuclear security.

(b) Terms that are used in both safety standards and nuclear security guidance with different meanings (and therefore different definitions). For example, the term *detection* in safety standards refers specifically to the detection of radiation by an instrument, whereas in security, the function of detection means detecting a nuclear security event, which involves: firstly, detecting (with an instrument) radiation from radioactive material, or detecting (with an instrument) other material that may be indicative of the presence of nuclear or other radioactive material, or receiving an information alert indicating the possibility of a criminal or intentional unauthorized act involving or directed against nuclear or other radioactive material or an associated facility or activity; and secondly, confirming by analysis that there is a nuclear security event.

USE OF THE IAEA NUCLEAR SAFETY AND SECURITY GLOSSARY

Interpretation of entries in the IAEA Nuclear Safety and Security Glossary

The entry for each term generally starts with one or more recommended definition(s).[9] Alternative definitions are given:

(a) If the term is used in two or more distinct nuclear safety or nuclear security related contexts (e.g. the term *clearance*, which is used for an administrative mechanism for removing regulatory control from material and for a biological process affecting the movement of inhaled radionuclides in the body);

(b) If it is necessary to include in the IAEA Nuclear Safety and Security Glossary an established definition that is not considered suitable as a general definition (in particular, some of the definitions from SSR-6 (Rev. 1) [2] and INFCIRC/225 [6] that may need to be retained in supporting publications but which are not the preferred general definitions);

(c) To include definitions of which drafters and reviewers of IAEA publications should be aware, even though they are unlikely to be used in IAEA publications (definitions in the main safety and security related conventions are an important example of this group); or

(d) For a small number of basic terms that have two distinct definitions, depending on whether they are being used in a scientific or regulatory context. An important example in the context of protection and safety is the adjective *radioactive*. Scientifically, something is described as *radioactive* if it exhibits the phenomenon of radioactivity or — in the somewhat less precise, but generally accepted, usage — if it contains any substance that exhibits radioactivity. Scientifically, therefore, virtually any material (including material that is considered to be waste) is radioactive. However, it is common regulatory practice to define terms such as *radioactive material* and *radioactive waste* in such a way as to include only that material or waste that is subject to regulation by virtue of the radiological hazard that it poses. Although the exact specifications vary from State to State, this typically excludes material and waste with very low concentrations of radionuclides and those that contain only 'natural' concentrations of naturally occurring radionuclides.

---

[9] A few terms are included without a recommended definition. In most such cases, the term in question is used to group a number of qualified terms and has no special meaning in itself (e.g. *emergency action level*, *recording level*, etc., are listed under *level*, but *level* itself is not defined). In some cases guidance is given on usage for terms with no definition.

Different definitions of a given term are numbered. Unless otherwise indicated in the text, drafters should use the definition that is most appropriate for their purposes.

In many cases, the definition is followed by a note as appropriate, such as:

(a) A note of caution (indicated by the symbol !), such as for terms that do not mean what they might appear to mean, or for potential conflicts with other safety or security related terminology

(b) A note of information (indicated by the symbol ①), such as:

- Explanation of the context(s) in which the term is usually used (and, in some cases, contexts in which it should not be used);

- Reference to related terms: synonyms, terms with similar but not identical meanings, 'contrasting' terms, and terms that supersede or are superseded by the term being described;

- Miscellaneous information, such as the units in which a quantity is normally measured, recommended parameter values and references.

(c) A special type of information note (indicated by the symbol §) to make the reader aware where there are terms or definitions in a safety context that might appear similar or related to terms in a nuclear security context, or vice versa.

This supplementary information is not part of the definition, but it is included to assist drafters and reviewers in understanding how to use (or how not to use) the term in question.

The use of *italics* in the text denotes a term or subterm with an entry in the IAEA Nuclear Safety and Security Glossary. The use of ***bold italics*** in the text denotes a subterm with its definition or with an explanation.

Use by drafters

Drafters of safety and security related IAEA publications — in particular safety standards and nuclear security guidance — should, as far as possible, use the terms in the IAEA Nuclear Safety and Security Glossary with the meanings given. Terms should also be used consistently. Every time a different term or form of words is used, the reader might question whether a different meaning is intended. Unnecessary variety of expression should be avoided if there is any possibility of causing confusion or ambiguity. Terms that are not listed in the IAEA Nuclear Safety and Security Glossary may be used, provided that there is no suitable alternative term listed in the IAEA Nuclear Safety and Security Glossary.

A new publication may contain a list of key terms used in that publication and their definitions. However, the first question concerning the inclusion of the definition of any term in a publication should always be whether the term actually needs to be defined. In other cases, terms should be explicitly defined in a publication only if a definition is essential to the correct understanding of that publication.

If a term is used with its normal dictionary meaning, or if its meaning in a particular publication will be obvious to the reader from the context, then there is no need for a definition. A term whose meaning is imprecise may need to be defined, if the imprecision actually detracts from a correct understanding of the text; in many cases, however, the precise meaning of a term will not matter for the purposes of a given publication. Similarly, obvious derivatives of a defined term need not themselves be defined unless there is some specific ambiguity that needs to be addressed.

If it is considered necessary to include a term in a list of definitions in an individual publication, the recommended, existing definition should be used wherever possible. If the recommended definition is not suitable (e.g. if the subject of the publication falls outside the scope of the existing definition), the wording of the definition may be modified, but its meaning should not be changed. The technical officer responsible for the IAEA Nuclear Safety and Security Glossary should be informed of any such modifications to the wording of definitions.

Similarly, definitions of any additional — usually more specialized — terms needed in a specific publication can be provided included either in the main text, in a footnote or in a list of definitions. Such definitions should be sent for information to the technical officer responsible for the IAEA Nuclear Safety and Security Glossary.

Some terms and usages that have been used in the past but are now obsolete, or that are used in the publications of other organizations but whose use is discouraged in IAEA publications, are included in the IAEA Nuclear Safety and Security Glossary. Such terms are listed in square brackets and should generally be avoided unless they are essential for referring to other publications; alternative terms for use in IAEA publications are provided. Similarly, some definitions are presented in square brackets, indicating that the term may be used but that the definition has been included for information only and should not be used as a working definition for IAEA publications.

The technical officer for a publication is responsible for ensuring that any definitions given in that publication are in accordance with these rules.

Terms defined in the IAEA Nuclear Safety and Security Glossary are likely to be used in informing the public on matters concerning nuclear safety and security and radiation risks, and in covering these matters in the news media. The technical terms used to explain difficult concepts will be interpreted and employed by writers, journalists and broadcasters who might not have a clear understanding of their significance. It must be borne in mind by drafters, reviewers and editors that certain terms that have specific and clear meanings in a scientific or technical context may be subject to misrepresentation or misunderstanding in a more general context.

The incautious use of language can and does give rise to widespread false impressions among the public that are difficult or impossible to correct. In attempting to summarize, interpret and simplify technical texts so as to communicate with a broader audience, therefore, care must be taken not to oversimplify by omitting conditions and qualifications, and not to mislead by using terms with both scientific and more general meanings.

Potentially misleading words include, for example, 'attributable', 'contamination', '[excess, statistical] deaths', 'exposure', 'illicit trafficking (in nuclear or radioactive material)', 'nuclear [terrorism, trafficking]', 'protection', 'radioactive', 'risk' and 'safe', and related words and phrases. Particular caution should be applied in relation to matters of life and health, especially fatal accidents and other major incidents, and to other emotionally charged subjects.

Finally, there are cases where words have taken on such a specific meaning in the IAEA context that the use of those words in their everyday sense could cause confusion. Examples include 'activity', 'critical', 'justification', 'practice', 'requirement', 'recommendation', 'guide' and 'standard' (and also 'shall' and 'should'). Although it would be unreasonable to prohibit the use of such words in their everyday sense in any IAEA publications, care should be taken to ensure that they are not used in a manner that could be ambiguous.

Use by reviewers

Reviewers should consider whether each term included in a list of definitions in an individual publication really needs to be defined, and if so whether the list of definitions (as opposed to the main text or a footnote) is the most appropriate place for the definition. Reviewers should also consider, of course, whether any terms not defined in the publication need to be defined.

If reviewers do not consider the above criteria to have been met, they should make appropriate recommendations to the IAEA technical officer responsible for the publication under review.

Reviewers should verify that defined terms and other words have been used in such a way that clear distinctions are drawn (or may be inferred) between, for example: events and situations (see the entry for *event*); accidents and other incidents; what is actual (i.e. what is), possible (i.e. what might be) or potential (i.e. what could become), and what is hypothetical (i.e. what is postulated or assumed); and what is observed or determined objectively and what is decided or declared subjectively.

Novel and revised terminology merits careful attention. The introduction of novel concepts and terminology can lead to difficulties in comprehension, and a profusion of newly defined terms can complicate drafting and review. Once terms have been defined, they should be used wherever applicable, and reviewers and reviewers will need to verify proper usage.

Electronic version of the IAEA Nuclear Safety and Security Glossary

The terms, definitions and notes in the IAEA Nuclear Safety and Security Glossary are available in an electronic version of the Glossary[10]. The electronic version allows quick access to the terms and definitions and also enables the user to navigate easily between broader, narrower and related terms.

The electronic version of the Glossary has been integrated with the IAEA Nuclear Safety and Security Online User Interface (NSS-OUI)[11], which is an electronic resource facilitating direct access to the content of the safety standards and nuclear security guidance and navigation within the texts. Terms contained in Safety Requirements publications within NSS-OUI are tagged with the appropriate definition, which can be displayed in a pop-up window. This helps to reduce any ambiguity between the choice of correct definition for readers of the publication.

FUTURE DEVELOPMENT OF THE IAEA NUCLEAR SAFETY AND SECURITY GLOSSARY

The IAEA Nuclear Safety and Security Glossary is intended to be reviewed and revised as necessary, to accurately represent the current terminology of the IAEA safety standards and nuclear security guidance. The review and revision of terminology is subject to appropriate consultation, as the IAEA Nuclear Safety and Security Glossary is also intended to bring about stability and harmonization in terminology and usage.

It is recognized that there are several terms and definitions, particularly in the area of nuclear security, where differences in usage still need to be reconciled and further harmonization is needed. The terms and definitions set out in this 2022 (Interim) Edition of the IAEA Nuclear Safety and Security Glossary may be used as a basis for discussions around achieving greater consistency of terminology. Any new terms and definitions developed within such a consultative process will be listed in the relevant publications and then included in future editions of the IAEA Nuclear Safety and Security Glossary.

Comments on the IAEA Nuclear Safety and Security Glossary may be provided by users of the IAEA safety standards and nuclear security guidance (in English and in translation) via the IAEA safety standards and nuclear security guidance contact point (Safety.Standards.Security.Guidance@iaea.org). Please read this Introduction before using the IAEA Nuclear Safety and Security Glossary and before submitting comments or queries.

---

[10] See https://kos.iaea.org/iaea-safety-glossary.html.

[11] See https://nucleus-apps.iaea.org/nss-oui/.

A

A_1

The *activity* value of *special form radioactive material* that is listed in Table 2 or derived in Section IV [both of the Transport Regulations] and is used to determine the *activity limits* for the *requirements* of [the Transport] Regulations. (See SSR-6 (Rev. 1) [2], sections II and IV and table 2.)

ⓘ A_1 is the maximum *activity* of *special form radioactive material* that can be transported in a Type A *package*. Fractions and multiples of A_1 are also used as criteria for other *package* types, etc.

ⓘ The corresponding value for any other form of *radioactive material* is A_2.

A_2

The *activity* value of *radioactive material*, other than *special form radioactive material*, that is listed in Table 2 or derived in Section IV [both of the Transport Regulations] and is used to determine the *activity limits* for the *requirements* of [the Transport] Regulations. (See SSR-6 (Rev. 1) [2], sections II and IV and table 2.)

ⓘ A_2 is the maximum *activity* of any *radioactive material* other than *special form radioactive material* that can be transported in a Type A *package*. Fractions and multiples of A_2 are also used as criteria for other *package* types, etc.

ⓘ The corresponding value for *special form radioactive material* is A_1.

abnormal operation

See *plant states (considered in design)*: *anticipated operational occurrence*.

absorbed dose

See *dose quantities*.

absorbed fraction

The fraction of energy emitted as a specified *radiation* type in a specified *source region* that is absorbed in a specified *target tissue*.

absorption

1.   See *sorption*.

2.   See *lung absorption type*.

absorption type, lung

See *lung absorption type*.

accelerogram

A recording of ground acceleration, usually in three orthogonal directions (i.e. components), two in the horizontal plane and one in the vertical plane.

acceptable limit

See *limit*.

acceptance criteria

Specified bounds on the value of a *functional indicator* or *condition indicator* used to assess the ability of a *structure, system or component* to perform its *design* function.

access delay

The element of a *physical protection system* designed to increase *adversary penetration time* for entry into and/or exit from the *nuclear facility* or *transport*.

> ⓘ *Access delay* can be accomplished by *physical barriers*, activated delays, complexity and/or personnel.

> ! Note that this is not the whole delay to which an adversary is subject, as it excludes the time needed to complete a *malicious act* after reaching the *target*.

accident

Any unintended *event*, including operating errors, equipment *failures* and other mishaps, the consequences or potential consequences of which are not negligible from the point of view of *protection and safety*.

> **criticality accident.** An *accident* involving *criticality*.

> ⓘ Typically, a *criticality accident* is an accidental release of energy as a result of unintentionally producing a *criticality* in a *facility* in which *fissile material* is used.

> ⓘ A *criticality accident* is also possible for *fissile material* in *storage* or in *transport*, for example.

> **nuclear accident.** [Any *accident* involving *facilities or activities* from which a *release of radioactive material* occurs or is likely to occur and which has resulted or may result in an international significant transboundary release that could be of radiological *safety* significance for another State.] (See Ref. [15].)

> ! This is not explicitly stated to be a definition of *nuclear accident*, but it is derived from the statement of the scope of application in Article 1 of the Convention on Early Notification of a Nuclear Accident. However, this Convention has a limited scope of application, and it is unreasonable to consider a *nuclear accident* to be only an *accident* that results or may result in an international significant transboundary release.

> **severe accident.** *Accident* more severe than a *design basis accident* and involving significant core degradation.

> ! In the 2008 INES Manual [16], there was a fundamental mismatch between the terminology used in *safety standards* and the designations used in *INES*. In short, *events* that would be considered *accidents* according to the *safety standards* definition may be *accidents* or *incidents* (i.e. not *accidents*) in *INES* terminology. This was not a serious day to day problem because the two areas are quite separate and have quite different purposes. However, it was a potential cause of confusion in communication with the news media and the public.

accident conditions

See *plant states (considered in design)*.

accident management

The taking of a set of actions during the evolution of an *accident*:

> (a) To prevent escalation to a *severe accident*;
>
> (b) To mitigate the consequences of a *severe accident*;
>
> (c) To achieve a long term safe stable state.

> ⓘ Aspect (b) of *accident management* (to mitigate the consequences of a *severe accident*) is also termed **severe accident management**.

> ⓘ By extension, *accident management* for a *severe accident* includes the taking of a set of actions during the evolution of the *accident* to mitigate degradation of the reactor core.

accident precursor

An *initiating event* that could lead to *accident conditions*.

accounting and control

See *system for nuclear material accounting and control*.

accuracy

See *validation* (1): *system code validation*.

activation

The *process* of inducing *radioactivity* in matter by irradiation of that matter.

> ⓘ In the context of nuclear installations, *activation* is used to refer to the unintentional induction of *radioactivity* in moderators, coolants, and structural and shielding materials, caused by irradiation with neutrons.

> ⓘ In the context of the production of radioisotopes, *activation* is used to refer to the intentional induction of *radioactivity* by neutron *activation*.

> ⓘ In other contexts, *activation* is an incidental side effect of irradiation carried out for other purposes, such as the sterilization of medical products or enhancement of the colour of gemstones for aesthetic reasons.

> ! Care may be needed to avoid confusion when using the term *activation* in its everyday sense of bringing into action (e.g. of *safety systems*, for which 'actuation' may be used).

activation product

A radionuclide produced by *activation*.

> ⓘ Often used to distinguish from *fission products*. For example, in *decommissioning waste* comprising structural materials from a *nuclear facility, activation products* might typically be found primarily within the matrix of the material, whereas *fission products* are more likely to be present in the form of *contamination* on surfaces.

active component

A *component* whose functioning depends on an external input such as actuation, mechanical movement or supply of power.

    ⓘ An *active component* is any *component* that is not a *passive component*.

    ⓘ Examples of *active components* are pumps, fans, relays and transistors. It is emphasized that this definition is necessarily general in nature, as is the corresponding definition of *passive component*. Certain *components*, such as rupture discs, check valves, *safety* valves, injectors and some solid state electronic devices, have characteristics that require special consideration before designation as an *active component* or a *passive component*.

    ⓘ Contrasting term: *passive component*.

See also *component*, *core components* and *structures, systems and components*.

    ! Care may be needed to avoid confusion with *radioactive components*.

activity

1.   The quantity A for an amount of radionuclide in a given energy state at a given time, defined as:

$$A(t) = \frac{dN}{dt}$$

where dN is the expectation value of the number of spontaneous nuclear transformations from the given energy state in the time interval dt.

    ⓘ   The rate at which nuclear transformations occur in a *radioactive material*. The equation is sometimes given as:

$$A(t) = -\frac{dN}{dt}$$

where N is the number of nuclei of the radionuclide, and hence the rate of change of N with time is negative. Numerically, the two forms are identical.

    ⓘ The SI unit for activity is reciprocal second (s^{-1}), termed the *becquerel* (Bq).

    ⓘ Formerly expressed in *curies* (Ci); *activity* values may be given in Ci (with the equivalent in Bq in parentheses) if they are being quoted from a reference that uses Ci as the unit.

specific activity. Of a radionuclide, the *activity* per unit mass of that nuclide.

The *specific activity* of a material is the *activity* per unit mass or volume of the material in which the radionuclides are essentially uniformly distributed.

The *specific activity* of a material, for the purposes of the Transport Regulations, is the *activity* per unit mass of the material in which the radionuclides are essentially uniformly distributed. (See SSR-6 (Rev. 1) [2].)

    ⓘ The distinction in usage between *specific activity* and **activity concentration** is controversial. Some regard the terms as synonymous, and may favour one or the other (as above). ISO 921: 1997 [17] distinguishes between *specific activity* as the *activity* per unit mass and *activity concentration* as the *activity* per unit volume.

ⓘ Another common distinction is that *specific activity* is used (usually as *activity* per unit mass) with reference to a pure sample of a radionuclide or, less strictly, to cases where a radionuclide is intrinsically present in the material (e.g. ^{14}C in organic materials, ^{235}U in *natural uranium*), even if the abundance of the radionuclide is artificially changed. In this usage, *activity concentration* (which may be *activity* per unit mass or per unit volume) is used for any other situation (e.g. when the *activity* is in the form of *contamination* in or on a material).

ⓘ In general, the term *activity concentration* is more widely applicable, is more self-evident in meaning, and is less likely than *specific activity* to be confused with unrelated terms (such as 'specified activities'). *Activity concentration* is therefore preferred to *specific activity* for general use in *safety* related *IAEA publications*.

2. See *facilities and activities*.

activity concentration

See *activity* (1): *specific activity*.

activity median aerodynamic diameter (AMAD)

The value of *aerodynamic diameter* such that 50% of the airborne *activity* in a specified aerosol is associated with particles smaller than the *AMAD*, and 50% of the *activity* is associated with particles larger than the *AMAD*.

ⓘ Used in internal dosimetry for simplification as a single 'average' value of *aerodynamic diameter* representative of the aerosol as a whole.

ⓘ The *AMAD* is used for particle sizes for which deposition depends principally on inertial impaction and sedimentation (i.e. typically those greater than about 0.5 μm).

activity median thermodynamic diameter (AMTD). For smaller particles, deposition typically depends primarily on *diffusion*, and the *AMTD* — defined in an analogous way to the *AMAD*, but with reference to the *thermodynamic diameter* of the particles — is used.

aerodynamic diameter. The *aerodynamic diameter* of an airborne particle is the diameter that a sphere of unit density would need to have in order to have the same terminal velocity when settling in air as the particle of interest.

thermodynamic diameter. The *thermodynamic diameter* of an airborne particle is the diameter that a sphere of unit density would need to have in order to have the same *diffusion* coefficient in air as the particle of interest.

activity median thermodynamic diameter (AMTD)

See *activity median aerodynamic diameter (AMAD)*.

actuated equipment

An assembly of *prime movers* and *driven equipment* used to accomplish one or more *safety tasks*.

actuation device

See *device*.

acute exposure

See *exposure*.

acute intake

See *intake* (2).

additive risk projection model

See *model*: *risk projection model*.

adsorption

See *sorption*.

advection

The movement of a substance or the transfer of heat by the motion of the gas (usually air) or liquid (usually water) in which it is present.

    ⓘ Sometimes used with the more common meaning — transfer of heat by the horizontal motion of the air — but in *IAEA publications* is more often used in a more general sense, in particular in *safety assessment*, to describe the movement of a radionuclide due to the movement of the liquid in which it is dissolved or suspended.

    ⓘ Usually contrasted with *diffusion*, where the radionuclide moves relative to the carrying medium.

adversary

Any individual performing or attempting to perform a *malicious act*.

    !   Where the term *threat* is used in the specific sense of an individual or group of individuals, an adversary is a person or group actually attempting to carry out a *malicious act*, whereas a *threat* is a postulated *adversary* against whom security measures are designed.

    external adversary. An *adversary* other than an *insider*.

    insider adversary. An *insider* that commits malicious activities with awareness, intent and motivation.

    See also *insider* and *threat*.

aerodynamic diameter

See *activity median aerodynamic diameter (AMAD)*.

aerodynamic dispersion

See *dispersion*.

ageing

General *process* in which characteristics of a *structure, system or component* gradually change with time or use.

14

ⓘ Although the term *ageing* is defined in a neutral sense — the changes involved in *ageing* may have no effect on *protection or safety*, or could even have a beneficial effect — it is most commonly used with a connotation of changes that are (or could be) detrimental to *protection and safety* (i.e. as a synonym of *ageing degradation*).

non-physical ageing. The *process* of becoming out of date (i.e. obsolete) owing to the evolution of knowledge and technology and associated changes in codes and standards.

ⓘ Examples of *non-physical ageing* effects include the lack of an effective *containment* or *emergency* core cooling *system*, the lack of *safety design* features (such as *diversity*, separation or *redundancy*), the unavailability of qualified spare parts for old equipment, incompatibility between old and new equipment, and outdated *procedures* or documentation (e.g. which thus do not comply with current regulations).

ⓘ Strictly, this is not always *ageing* as defined above, because it is sometimes not due to changes in the *structure, system or component* itself. Nevertheless, the effects on *protection and safety*, and the solutions that need to be adopted, are often very similar to those for *physical ageing*.

ⓘ The term **technological obsolescence** is also used.

physical ageing. *Ageing* of *structures, systems and components* due to physical, chemical and/or biological *processes* (*ageing* mechanisms).

ⓘ Examples of *ageing* mechanisms include wear, thermal or *radiation* embrittlement, corrosion and microbiological fouling.

ⓘ The term **material ageing** is also used.

accelerated ageing. A method of equipment testing in which the *ageing* associated with longer term service conditions is simulated in a short time.

ⓘ Usually, accelerated ageing attempts to simulate natural *ageing* effects by application of stressors representing pre-service and service conditions, but with differences in intensity, duration and the manner of application.

ageing degradation

Ageing effects that could impair the ability of a *structure, system or component* to function within its *acceptance criteria*.

ⓘ Examples include reduction in diameter due to wear of a rotating shaft, loss in material toughness due to *radiation* embrittlement or thermal *ageing*, and cracking of a material due to fatigue or stress corrosion cracking.

ageing management

Engineering, *operations* and *maintenance* actions to control within *acceptable limits* the *ageing degradation* of *structures, systems and components*.

ⓘ Examples of engineering actions include *design, qualification* and *failure analysis*. Examples of *operations* actions include *surveillance*, carrying out operating *procedures* within specified *limits* and performing environmental measurements.

life management (or **lifetime management**). The integration of *ageing management* with economic planning: (1) to optimize the *operation, maintenance* and *service life* of *structures, systems and components*; (2) to maintain an acceptable level of *safety* and performance; and (3) to improve economic performance over the *service life* of the *facility*.

ageing mechanism

A process that gradually changes the characteristics of a structure, *system or component* over time or with use (e.g. curing, wear, fatigue, creep, erosion, microbiological fouling, corrosion, embrittlement, chemical decomposition).

> **significant ageing mechanism.** An *ageing mechanism* that under normal and abnormal service conditions causes degradation of equipment that makes the equipment vulnerable to failure to perform its *safety function* in *accident conditions*.

agricultural countermeasure

See *countermeasure*.

air kerma

See *kerma*.

aircraft

> **cargo aircraft.** Any *aircraft*, other than a *passenger aircraft*, that is carrying goods or property. (See SSR-6 (Rev. 1) [2].)

> **passenger aircraft.** An *aircraft* that carries any person other than a crew member, a *carrier*'s employee in an official capacity, an authorized representative of an appropriate national authority, or a person accompanying a consignment or other cargo. (See SSR-6 (Rev. 1) [2].)

ALARA (as low as reasonably achievable)

See *optimization (of protection and safety)*.

alarm

> **false alarm.** An alarm found by subsequent assessment not to have been caused by the presence of *nuclear or radioactive material*.

> ⓘ   This definition is for use in relation to detection of material *out of regulatory control*.

> **innocent alarm.** An alarm found by subsequent assessment to have been caused by *nuclear or other radioactive material* under *regulatory control* or exempt or excluded from *regulatory control*.

> ⓘ   It is therefore a valid alarm: the system indicated the presence of material as it was designed to do, but the subsequent analysis showed that the material was not of security concern. This definition is for use in relation to detection of material *out of regulatory control*.

> **instrument alarm.** Signal from instruments that could indicate a *nuclear security event*, requiring assessment. An *instrument alarm* may come from devices that are portable or deployed at fixed locations and operated to augment normal commerce protocols and/or in a law enforcement operation.

> ⓘ   This definition is for use in the context of *nuclear security*.

nuisance alarm. A *false alarm* or an *innocent alarm*.

ⓘ This definition is for use in relation to detection of material *out of regulatory control*.

aleatory uncertainty

See *uncertainty*.

alternate AC power source

A power source reserved for the use for the power supply to the plant during total loss of all non-battery power in the safety power systems (*station blackout*) and other *design extension conditions*.

alert

See *emergency class*.

ambient dose equivalent

See *dose equivalent quantities (operational)*.

analysis

ⓘ Often used interchangeably with *assessment*, especially in more specific terms such as '*safety analysis*'. In general, however, *analysis* suggests the *process* and result of a study aimed at understanding the subject of the *analysis*, while *assessment* may also include determinations or judgements of acceptability. *Analysis* is also often associated with the use of a specific technique. Hence, one or more forms of *analysis* may be used in *assessment*.

cost–benefit analysis. A systematic technical and economic evaluation of the positive effects (benefits) and negative effects (disbenefits, including monetary costs) of undertaking an action.

ⓘ A decision aiding technique commonly used in the *optimization of protection and safety*. This and other techniques are discussed in Ref. [18].

event tree analysis. An inductive technique that starts by hypothesizing the occurrence of basic *postulated initiating events* and proceeds through their logical propagation to *system failure events*.

ⓘ The *event* tree is the diagrammatic illustration of alternative outcomes of specified *postulated initiating events*.

ⓘ *Fault tree analysis* considers similar chains of *events*, but starts at the other end (i.e. with the 'results' rather than the 'causes'). The completed *event* trees and fault trees for a given set of *events* would be similar to one another.

fault tree analysis. A deductive technique that starts by hypothesizing and defining *failure events* and systematically deduces the *events* or combinations of *events* that caused the *failure events* to occur.

ⓘ The fault tree is the diagrammatic illustration of the *events*.

ⓘ *Event tree analysis* considers similar chains of *events*, but starts at the other end (i.e. with the 'causes' rather than the 'results'). The completed *event* trees and fault trees for a given set of *events* would be similar to one another.

hazard analysis. A process of examining a system throughout its life cycle to identify inherent *hazards* and *contributory hazards*, and requirements and constraints to eliminate, prevent or control them.

ⓘ The scope of *hazard analysis* extends beyond *design basis accidents* for the plant by including abnormal events and plant operations with degraded equipment and plant systems.

safety analysis. Evaluation of the potential *hazards* associated with the operation of a *facility* or the conduct of an *activity*.

ⓘ The formal *safety analysis* is part of the overall *safety assessment*; that is, it is part of the systematic process that is carried out throughout the design process (and throughout the *lifetime* of the *facility* or the *activity*) to ensure that all the relevant *safety requirements* are met by the proposed (or actual) design.

ⓘ *Safety analysis* is often used interchangeably with *safety assessment*. However, when the distinction is important, *safety analysis* should be used as a documented process for the study of *safety*, and *safety assessment* should be used as a documented process for the evaluation of *safety* — for example, evaluation of the magnitude of hazards, evaluation of the performance of *safety measures* and judgement of their adequacy, or quantification of the overall radiological impact or *safety* of a *facility* or *activity*.

sensitivity analysis. A quantitative *examination* of how the behaviour of a *system* varies with change, usually in the values of the governing parameters.

ⓘ A common approach is parameter variation, in which the variation of results is investigated for changes in the value of one or more input parameters within a reasonable range around selected reference or mean values, and perturbation *analysis*, in which the variations of results with respect to changes in the values of all the input parameters are obtained by applying differential or integral *analysis*.

static analysis. *Analysis* of a system or component based upon its form, structure, content or documentation.

uncertainty analysis. An *analysis* to estimate the uncertainties and error bounds of the quantities involved in, and the results from, the solution of a problem.

annual dose

See *dose concepts*.

annual limit on exposure (ALE)

See *limit*.

annual limit on intake (ALI)

See *limit*.

annual risk

See *risk* (3).

anticipated operational occurrence

See *plant states (considered in design)*.

anticipated transient without scram (ATWS)

For a nuclear reactor, an *accident* for which the *initiating event* is an *anticipated operational occurrence* and in which the *system* for fast *shutdown* of the reactor fails to function.

applicant

Any *person or organization* applying to a *regulatory body* for *authorization* (or *approval*) to undertake specified *activities*.

ⓘ Strictly, an *applicant* would be such from the time at which an application is submitted until the requested *authorization* is either granted or refused. However, the term is often used a little more loosely than this, in particular in cases where the *authorization process* is long and complex.

approval

The granting of consent by a *regulatory body.*

ⓘ Typically used to represent any form of consent from the *regulatory body* that does not meet the definition of *authorization*. In the context of the Transport Regulations [2], *approval* is a specific type of *authorization* concerning the items listed in para. 802 of the Transport Regulations that is carried out in accordance with the applicable requirements of the Transport Regulations.

multilateral approval. *Approval* by the relevant *competent authority* of the country of origin of the design or *shipment*, as applicable, and also, where the *consignment* is to be transported *through* or *into* any other country, *approval* by the *competent authority* of that country. (See SSR-6 (Rev. 1) [2].)

unilateral approval. An *approval* of a *design* that is required to be given by the *competent authority* of the country of origin of the *design* only. (See SSR-6 (Rev. 1) [2].)

architecture

1. Organizational structure of the instrumentation and *control systems* of a plant that are important to safety.

2. See *nuclear security detection architecture.*

area

controlled area. A defined area in which specific *protection* measures and *safety* provisions are or could be required for controlling *exposures* or preventing the spread of *contamination* in normal working conditions, and preventing or limiting the extent of *potential exposures.*

ⓘ A *controlled area* is often within a *supervised area*, but need not be.

ⓘ The term [*radiation area*] is sometimes used to describe a similar concept, but *controlled area* is preferred in *IAEA publications.*

hazard control area. A designated geographical area, representing the maximum extent of all hazards within a *radiological crime scene*, into which, within and from which access is controlled.

inner area. An area with additional protection measures inside a *protected area*, where *Category I nuclear material* is used and/or stored.

limited access area. Designated area containing a *nuclear facility* and *nuclear material* to which access is limited and controlled for physical protection purposes.

operational control area. A designated geographical area, representing the maximum extent of the area needed to support the management of a *radiological crime scene*, into and from which access is controlled.

operations area. A geographical area that contains an *authorized facility*. It is enclosed by a physical *barrier* (the **operations boundary**) to prevent unauthorized access, by means of which the management of the *authorized facility* can exercise direct authority.

ⓘ This applies to larger *facilities*.

protected area. Area inside a *limited access area* containing *Category I or II nuclear material* and/or *sabotage targets* surrounded by a *physical barrier* with additional *physical protection measures*.

[*radiation area*]. See *area: controlled area*.

site area. A geographical area that contains an *authorized facility, authorized activity* or *source*, and within which the management of the *authorized facility* or *authorized activity* or first responders may directly initiate *emergency response actions*.

ⓘ This is typically the area within the security perimeter fence or other designated property marker. It may also be the *controlled area* around industrial radiography work or an inner cordoned off area established by *first responders* around a suspected *hazard*.

ⓘ The boundary of the *site area* is called the **site boundary**.

ⓘ This area is often identical to the *operations area*, except in situations (e.g. *research reactors, irradiation installations*) where the *authorized facility* is on a site where other *activities* are being carried out beyond the *operations area*, but where the management of the *authorized facility* can be given some degree of authority over the whole *site area*.

ⓘ The term *activity* is used here in the sense of *activity* (2).

supervised area. A defined area not designated as a *controlled area* but for which *occupational exposure* conditions are kept under review, even though specific *protection* measures or *safety* provisions are not normally needed.

See also *controlled area*.

vital area. Area inside a *protected area* containing equipment, systems or devices, *or nuclear material*, the *sabotage* of which could directly or indirectly lead to *high radiological consequences*.

area monitoring

See *monitoring* (1).

area survey

See *survey*.

arrangements (for emergency response)

See *emergency arrangements*.

arrangements (for operations)

The integrated set of infrastructural elements necessary to provide the capability for performing a specified function or task required to carry out a specified operation.

ⓘ The infrastructural elements may include authorities and responsibilities, organization, coordination, personnel, plans, *procedures*, *facilities*, equipment or training.

assessment

1.   The *process*, and the result, of analysing systematically and evaluating the hazards associated with *sources* and *practices*, and associated *protection and safety* measures.

ⓘ *Assessment* is often aimed at quantifying performance measures for comparison with criteria.

ⓘ In *IAEA publications*, *assessment* should be distinguished from *analysis*. *Assessment* is aimed at providing information that forms the basis of a decision on whether or not something is satisfactory. Various kinds of *analysis* may be used as tools in doing this. Hence an *assessment* may include a number of *analyses*.

consequence assessment. *Assessment* of the radiological consequences (e.g. *doses*, *activity concentrations*) of *normal operation* and possible *accidents* associated with an *authorized facility* or part thereof.

! Care should be taken in discussing 'consequences' in this context to distinguish between radiological consequences of events causing *exposure*, such as *doses*, and health consequences, such as cancers, that could result from *doses*. 'Consequences' of the former type generally imply a probability of experiencing 'consequences' of the latter type.

ⓘ This differs from *risk assessment* in that probabilities are not included in the *assessment*.

See also *end point*.

dose assessment. *Assessment* of the *dose(s)* to an individual or group of people.

ⓘ For example, *assessment* of the *dose* received or committed by an individual on the basis of results from *workplace monitoring* or *bioassay*.

ⓘ The term **exposure assessment** is also sometimes used.

hazard assessment. *Assessment* of *hazards* associated with *facilities*, *activities* or *sources* within or beyond the borders of a State in order to identify:

(a)   Those *events* and the associated areas for which *protective actions* and *other response actions* may be required within the State;

(b)   Actions that would be effective in mitigating the consequences of such *events*.

performance assessment. *Assessment* of the performance of a *system* or subsystem and its implications for *protection and safety* at an *authorized facility*.

ⓘ This differs from *safety assessment* in that it can be applied to parts of an *authorized facility* (and its surroundings) and does not necessarily require the *assessment* of radiological impacts.

See also *performance testing*.

radiological environmental impact assessment. *Assessment* of the expected radiological impacts of *facilities and activities* on the *environment* for the purposes of protection of the public and *protection of the environment* against *radiation risks*.

risk assessment.

1. *Assessment* of the *radiation risks* and other risks associated with *normal operation* and possible *accidents* involving *facilities and activities*.

ⓘ This will normally include *consequence assessment*, together with some *assessment* of the probability of those consequences arising.

2. The overall process of systematically identifying, estimating, analysing and evaluating *risk* for the purpose of informing priorities, developing or comparing courses of action, and informing decision making.

ⓘ This definition is broader than definition (1).

safety assessment.

1. *Assessment* of all aspects of a *practice* that are relevant to *protection and safety*; for an *authorized facility*, this includes *siting*, *design* and *operation* of the *facility*.

ⓘ This will normally include *risk assessment (1)*.

See also *probabilistic safety assessment (PSA)*.

2. *Analysis* to predict the performance of an overall *system* and its impact, where the performance measure is the radiological impact or some other global measure of the impact on *safety*.

3. The systematic process that is carried out throughout the design process (and throughout the *lifetime* of the *facility* or the *activity*) to ensure that all the relevant *safety requirements* are met by the proposed (or actual) design.

ⓘ *Safety assessment* includes, but is not limited to, the formal *safety analysis*; that is, it includes the evaluation of the potential hazards associated with the operation of a *facility* or the conduct of an *activity*.

ⓘ Stages in the *lifetime* of a *facility* or *activity* at which a *safety assessment* is carried out and updated and the results are used by the designers, the *operating organization* and the *regulatory body* include:

    (a) *Site evaluation* for the *facility* or *activity*;

    (b) Development of the *design*;

    (c) Construction of the *facility* or implementation of the *activity*;

    (d) Commissioning of the *facility* or of the *activity*;

    (e) Commencement of *operation* of the *facility* or conduct of the *activity*;

    (f) *Normal operation* of the *facility* or normal conduct of the *activity*;

    (g) Modification of the *design* or *operation*;

    (h) *Periodic safety reviews*;

    (i) Life extension of the *facility* beyond its original *design life*;

    (j) Changes in ownership or management of the *facility*;

    (k) *Decommissioning* of a *facility*;

    (l) *Closure* of a *disposal facility* for *radioactive waste* and the post-*closure* phase;

    (m) *Remediation* of a site and *release* from *regulatory control*.

See GSR Part 4 (Rev. 1) [19].

threat assessment. An evaluation of the *threats* — based on available intelligence, law enforcement, and open source information — that describes the motivation, intentions, and capabilities of these *threats*.

vulnerability assessment. A process that evaluates and documents the features and effectiveness of the overall security system at a particular *target*.

2. *Activities* carried out to determine whether *requirements* are met and *processes* are adequate and effective, and to encourage managers to implement improvements, including *safety* improvements.

 ⓘ This usage originated in *quality assurance* and related fields.

 ! The IAEA revised the *requirements* and guidance in the subject area of *quality assurance* for *safety standards* on *management systems* for the *safety* of *facilities and activities* involving the use of *ionizing radiation*. The terms quality management and *management system* have been adopted in the revised standards in place of the terms *quality assurance* and *quality assurance* programme.

 ⓘ *Assessment activities* may include reviewing, checking, inspecting, testing, surveillance, auditing, peer evaluation and technical review. These *activities* can be divided into two broad categories: *independent assessment* and *self-assessment*.

independent assessment. *Assessments* such as *audits* or surveillance carried out to determine the extent to which the *requirements* for the *management system* are fulfilled, to evaluate the effectiveness of the *management system* and to identify opportunities for improvement. They can be conducted by or on behalf of the organization itself for internal purposes, by interested parties such as customers and regulators (or by other persons on their behalf), or by external independent organizations.

 ⓘ This definition applies in *management systems* and related fields.

 ⓘ Persons conducting *independent assessments* do not participate directly in the work being assessed.

 ⓘ *Independent assessment activities* include internal and external *audit*, surveillance, peer evaluation and technical review, which are focused on *safety* aspects and areas where problems have been found.

 ⓘ An ***audit*** is used in the sense of a documented activity performed to determine by investigation, *examination* and evaluation of objective evidence the adequacy of, and adherence to, established *procedures*, instructions, specifications, codes, standards, administrative or operational programmes and other applicable documents, and the effectiveness of their implementation.

self-assessment. A routine and continuing *process* conducted by *senior management* and also by management at other levels to evaluate the effectiveness of performance in all areas of their responsibility.

 ⓘ This definition applies in *management systems* and related fields.

 ⓘ *Self-assessment activities* include review, surveillance and discrete checks, which are focused on preventing, or identifying and correcting, management problems that hinder the achievement of the organization's objectives, in particular *safety* objectives.

 ⓘ *Self-assessment* provides an overall view of the performance of the organization and the degree of maturity of the *management system*. It also helps to identify areas for improvement in the organization, to determine priorities and to set a baseline for further improvement.

See also *senior management*.

assisted operation

An operation undertaken by a State or group of States to which assistance is provided by or through the IAEA in the form of materials, services, equipment, *facilities* or information pursuant to an agreement between the IAEA and that State or group of States.

    ⓘ The word 'operation' is used here in its usual sense.

associated activity

See *facilities and activities*.

associated facility

See *facilities and activities*.

atmospheric dispersion

See *dispersion*.

attack scenario

See *scenario*.

attenuation

The reduction in intensity of *radiation* passing through matter due to *processes* such as *absorption* and scattering.

    ⓘ By analogy, also used in other situations in which some radiological property, characteristic or parameter is gradually reduced in the course of passing through a medium (e.g. the reduction in *activity concentration* in groundwater passing through the *geosphere* due to *processes* such as *sorption*).

attributable risk

See *risk* (3).

audit

See *assessment* (2): *independent assessment*.

authorization

The granting by a *regulatory body* or other governmental body of written permission for a *person or organization* (the *operator*) to conduct specified *activities*.

    ⓘ *Authorization* could include, for example, licensing (issuing a *licence*), *certification* (issuing a *certificate*) or *registration*.

    ⓘ The term *authorization* is also sometimes used to describe the document granting such permission.

    ⓘ In nuclear security, the governmental body concerned will be a *competent authority*, and the definition is set out in Nuclear Security Series publications as "The granting by a *competent authority* of written permission for operation of an *associated facility* or for carrying out an *associated activity* [or a document granting such permission]" [7, 8, 9].

ⓘ *Authorization* is generally a more formal *process* than *approval*. In the context of the Transport Regulations [2], *approval* is a specific type of *authorization* concerning the items listed in para. 802 of the Transport Regulations that is carried out in accordance with the applicable requirements of the Transport Regulations.

See also *approval*: *multilateral approval* and *unilateral approval.*

authorized activity

See *facilities and activities*.

authorized discharge

See *discharge* (1).

authorized facility

See *facilities and activities*.

authorized limit

See *limit*.

authorized party

The *person or organization* (the *operator*) responsible for an *authorized facility* or an *authorized activity* that gives rise to *radiation risks* who has been granted written permission (i.e. authorized) by a *regulatory body* or other governmental body to conduct specified activities.

ⓘ The *authorized party* for an authorized *facility* or an authorized activity is usually the *operating organization* or the *registrant* or *licensee* (although forms of *authorization* other than *registration* or licensing may apply) [13].

ⓘ In Nuclear Security Series publications, the term **authorized person** has been defined with a similar meaning, i.e. "A natural or legal person that has been granted an *authorization*. An *authorized person* is often referred to as a 'licensee' or '*operator*'" [7, 8, 9].

! *Authorized person* is sometimes used to refer an individual who has been authorized to carried out specific tasks or operate specific equipment. If used in this way, particular care should be taken to ensure that there is no possibility of confusion.

authorized termination of responsibility

The *release* by the *regulatory body* of an *operator* (or a former *operator*) from any further regulatory responsibilities in relation to an *authorized facility* or *authorized activity*.

ⓘ This may be a separate *process* from termination of an *authorization*; for example, termination of the responsibility to maintain active *institutional control* over a *disposal facility* or termination of the *authorization* for *decommissioning*.

authorized transfer

The transfer of regulatory responsibility for specified *radioactive material* from one *operator* to another.

> ! This does not necessarily involve any movement of the material itself.

authorized use

See *use*.

availability

1. The ability of an item to be in a state to perform a required function under given conditions at a given instant of time or over a given time interval, with the assumption that the necessary external resources are provided.

> (i) In probabilistic safety assessment, availability of an item is the probability that this item is capable of performing its required function at a certain point in time.

> (i) The definition was previously "The fraction of time for which a *system* is capable of fulfilling its intended purpose".

2. The property of being accessible and usable upon demand by an authorized entity.

> (i) This definition is for use in the context of security of nuclear information.

averted dose

See *dose concepts*.

B

backfill

Material used to refill excavated portions of a *disposal facility* after *waste* has been emplaced.

background

The *dose* or *dose rate* (or an observed measure related to the *dose* or *dose rate*) attributable to all *sources* other than the one(s) specified.

ⓘ Strictly, this applies to measurements of *dose rate* or count rate from a sample, where the *background dose rate* or count rate must be subtracted from all measurements. However, *background* is used more generally, in any situation in which a particular *source* (or group of *sources*) is under consideration, to refer to the effects of other *sources*. It is also applied to quantities other than *doses* or *dose rates*, such as *activity concentrations* in environmental media.

natural background. The *doses, dose rates* or *activity concentrations* associated with *natural sources,* or any other *sources* in the *environment* that are not amenable to *control*.

ⓘ This is normally considered to include *doses, dose rates* or *activity concentrations* associated with *natural sources* and global fallout (but not local fallout) from atmospheric nuclear weapon tests.

barrier

A physical obstruction that prevents or inhibits the movement of people, radionuclides or some other phenomenon (e.g. fire), or provides shielding against *radiation*.

See also *cladding, containment* (1) and *defence in depth* (1).

fire barrier. Wall, floor, ceiling or device for closing a passage such as a door, a hatch, a penetration or a ventilation system to limit the consequences of a fire.

ⓘ A *fire barrier* is characterized by a fire resistance rating.

intrusion barrier. *Components* of a *disposal facility* designed to prevent inadvertent access to the *waste* by people, animals or plants.

multiple barriers. Two or more natural or engineered *barriers* used to isolate *radioactive waste* in, and to prevent or to inhibit *migration* of radionuclides from, a *disposal facility*.

! The term 'chemical *barrier*' is sometimes used in the context of *waste disposal* to describe the chemical effect of a material that enhances the extent to which radionuclides react chemically with the material or with the host rock, thus inhibiting the *migration* of the radionuclides.

ⓘ This is not strictly a *barrier* as defined above (unless the material also constitutes a physical *barrier*), but the effect may be equivalent to that of a *barrier*, and it may therefore be convenient to regard it as such.

multiple safety functions. In the context of the fulfilment of *multiple safety functions* by a *disposal system*, the *containment* and *isolation* of *waste* (the *confinement* function) is fulfilled by two or more natural or engineered *barriers* of the *disposal facility*, by means of diverse physical and chemical properties or processes, together with operational controls.

physical barrier. A fence, wall or similar impediment which provides *access delay* and complements access control.

ⓘ This definition is for use in the context of *physical protection* of *nuclear material* and *nuclear facilities* and other *radioactive material* and *associated facilities and activities*.

Bayesian statistics

ⓘ *Bayesian statistics* provide a means for probabilistic inference that depends on the specification of prior distributions for all unknown parameters, followed by an application of Bayes' theorem to incorporate the extra information included in the data.

ⓘ *Bayesian statistics* can be used in volcanology, for example, as a method to help constrain the results and uncertainty estimates of statistical and numerical modelling, by taking advantage of as much data and relevant information as are available. In contrast, frequentist statistics rely on patterns of past events to model the likelihood that an event will occur in the future.

ⓘ Bayesian methods can incorporate more geological information into an estimate of probability of occurrence than is possible with a frequentist approach.

becquerel (Bq)

The SI unit of *activity*, equal to one (transformation) per second.

ⓘ Supersedes the non-SI unit *curie (Ci)*. 1 Bq = 27 pCi (2.7×10^{-11} Ci) approximately. 1 Ci = 3.7×10^{10} Bq.

beyond design basis accident

See *plant states (considered in design)*.

beyond design basis earthquake

The seismic ground motion (represented by acceleration time history or ground motion response spectra) corresponding to an earthquake severity higher than the one used for design.

ⓘ It is derived from the hazard evaluation of the site and is used in seismic margin assessment or seismic probabilistic safety assessment.

bias

A measure of the systematic error between an actual or true value and a prediction by a *model* or a measured mean value. The *bias* of a model represents the tendency of a *model* to overpredict or to underpredict.

bioassay

Any *procedure* used to determine the nature, *activity*, location or retention of radionuclides in the body by direct (in vivo) measurement or by in vitro analysis of material excreted or otherwise removed from the body.

ⓘ Sometimes referred to as 'radio-bioassay'.

biological half-life

See *half-life (2)*.

biophysical model

See *model*.

biosphere

That part of the *environment* normally inhabited by living organisms.

> ⓘ In practice, the *biosphere* is not usually defined with great precision, but is generally taken to include the atmosphere and the Earth's surface, including the soil and surface water bodies, seas and oceans, and their sediments. There is no generally accepted definition of the depth below the surface at which soil or sediment ceases to be part of the *biosphere*, but this might typically be taken to be the depth affected by basic human activities, in particular, farming.

> ⓘ In the *safety* of *radioactive waste management*, in particular, the *biosphere* is normally distinguished from the *geosphere*.

blackout (station)

See *station blackout*.

blended attack

A *malicious act* involving the coordinated use of both *cyber-attack* and physical attack.

book inventory

See *inventory*.

buffer

Any substance placed around a *waste package* in a *disposal facility* to serve as a *barrier* to restrict the access of groundwater to the *waste package* and to reduce by *sorption* and precipitation the rate of eventual *migration* of radionuclides from the *waste*.

> ⓘ The above definition is clearly specific to the *safety* of *radioactive waste management*. The term buffer (e.g. in buffer solution) is also used, in its usual scientific sense (and therefore usually without specific definition), in a variety of contexts.

bulk analysis

The analysis of either an entire sample or a portion of the sample to determine the average properties of the measured portion.

burnable absorber

Neutron absorbing material, used to manage *reactivity*, with the particular capability of being depleted by neutron *absorption*.

> ⓘ A *burnable absorber* is used to manage *reactivity* by flattening the radial neutron flux within a reactor and to compensate for the depletion of *fissile material* due to operation of the reactor, thereby improving the utilization of the fuel.

[burnable poison]

See *burnable absorber* and *poison*.

bypass

1.   A device to inhibit, deliberately but temporarily, the functioning of a circuit or *system* by, for example, short circuiting the contacts of a relay.

> ***maintenance bypass.*** A *bypass* of *safety system* equipment during *maintenance*, testing or *repair*.

> ***operational bypass.*** A *bypass* of certain *protective actions* when they are not necessary in a particular mode of plant *operation*.

>> ! An *operational bypass* may be used when the *protective action* prevents, or might prevent, reliable *operation* in the required mode.

2.   A route that allows *fission products* released from a reactor core to enter the *environment* without passing through the *containment* or other enclosure designed to confine and reduce a *radioactive release* in the event of an *emergency*.

> ⓘ This route may be established intentionally by the *operator* or as a result of the *event*.

C

calibration

A set of *operations* that establish, under specified conditions, the relationship between values of quantities indicated by a measuring instrument or measuring system, or values represented by a material measure or a reference material, and the corresponding values realized by standards.

ⓘ The definition was previously "A measurement of, or adjustment to, an instrument, *component* or *system* to ensure that its accuracy or response is acceptable".

ⓘ A *calibration* may be expressed by a statement, *calibration* function, *calibration* diagram, *calibration* curve or *calibration* table. In some cases, it may consist of an additive or multiplicative correction of the indication with associated measurement uncertainty.

ⓘ *Calibration* should not be confused with adjustment of a measuring system, often mistakenly called 'self-calibration', or with *verification* of *calibration*.

calibration of a dosimeter. The *process* by which a dosimeter is characterized with a *calibration* factor. The *calibration* factor is the quotient of the conventionally true value of the measured quantity and the indicated value of the dosimeter under reference conditions. If the dosimeter is used under reference conditions, the value of the measured quantity is the product of the indicated value and the calibration factor. If the dosimeter is used under non-reference conditions, the value of the measured quantity is the product of the indicated value, the calibration factor and additional correction factor(s).

model calibration. The *process* whereby predictions by a *model* are compared with field observations and/or experimental measurements from the *system* being modelled, and the *model* is adjusted for *bias* if necessary to achieve a best fit to the measured and/or observed data.

! This usage of the term is not universally accepted. The terms *model validation* and *model verification* are more commonly used to describe related *processes* in relation to *models*.

See also *bias*.

calibration of a dosimeter

See *calibration*.

canister, waste

See *waste container*.

capable fault

See *geological fault*.

capable volcanic field

See *volcano: capable volcano*.

capable volcano

See *volcano*.

carers and comforters

Persons who willingly and voluntarily help (other than in their occupation) in the care, support and comfort of *patients* undergoing *radiological procedures* for medical diagnosis or medical treatment.

cargo aircraft

See *aircraft*.

carrier

Any person, organization or government undertaking the carriage of *radioactive material* by any means of *transport*.

> ⓘ The term includes both *carriers* for hire or reward (known as common or contract *carriers* in some countries) and *carriers* on own account (known as private *carriers* in some countries). (See SSR-6 (Rev. 1) [2].)

Category I/II/III nuclear material

See *nuclear material (1)*.

cause

> **direct cause.** The *latent weakness* (and the reasons for the *latent weakness*) that allows or causes the *observed cause* of an *initiating event* to happen.

> ⓘ Corrective actions designed to address *direct causes* are sometimes termed *repairs*.

> **latent weakness.** An undetected degradation in an element of a *safety layer*.

> ⓘ Such a degradation could lead to that element failing to perform as expected if it were called upon to perform a function.

> **observed cause.** The *failure*, action, omission or condition that directly leads to an *initiating event*.

> **root cause.** The fundamental cause of an *initiating event*, correction of which will prevent recurrence of the *initiating event* (i.e. the *root cause* is the *failure* to detect and correct the relevant *latent weakness(es)* and the reasons for that *failure*).

> ⓘ Corrective actions designed to address *root causes* are sometimes termed *remedies*.

central alarm station

An installation that provides for the complete and continuous alarm monitoring, assessment and communication with *guards*, facility management and *response forces*.

certificate

A legal document issued by the *regulatory body* stating the applicable conditions to be met for *certification* and certifying compliance with regulatory *requirements* if the conditions are met.

> ⓘ *Certificates* are required for some *package* types [2].

certification

The issue of a *certificate*.

channel

An arrangement of interconnected *components* within a *system* that initiates a single output.

> ⓘ A *channel* loses its identity where single output signals are combined with signals from other *channels* (e.g. from a monitoring *channel* or a *safety* actuation *channel*).

> ⓘ The above definition is specific to a particular area of *nuclear safety*. The term channel is also used in its usual senses (and therefore usually without specific definition) in a variety of contexts.

chain of custody

The procedures and documents that account for the integrity of physical evidence by tracking its handling and storage from its point of collection to its final disposition.

> ⓘ Other terms for this process are 'chain of evidence', 'chain of physical custody' and 'chain of possession'.

characterization

1. Determination of the nature and *activity* of radionuclides present in a specified place.

> ⓘ For example, *characterization* is the determination of the radionuclides present in a *bioassay* sample or in an area contaminated with *radioactive material* (e.g. as a first step in planning *remediation*). For the latter example, care should be taken to avoid confusion with the existing, and different, definition of the term *site characterization*.

2. Determination of the nature of the *radioactive material* and associated evidence.

> ⓘ This definition is for use in the context of *nuclear forensics*.

3. Determination of the character of something.

> ⓘ This is the standard dictionary definition and would not need to be included in an individual glossary. It is included here only to distinguish the usual usage from the more restricted usage indicated in (1) and (2).

characterization of waste. Determination of the physical, mechanical, chemical, radiological and biological properties of *radioactive waste* to establish the need for further adjustment, *treatment* or *conditioning*, or its suitability for further handling, *processing*, *storage* or *disposal*.

> ⓘ *Characterization of waste*, in accordance with *requirements* established or approved by the *regulatory body*, is a process in the *predisposal management* of *waste* that at various steps provides information relevant to process control and provides assurance that the *waste form* or *waste package* will meet the *waste acceptance criteria* for the *processing*, *storage*, *transport* and *disposal* of the *waste*.

site characterization (of the site for a *disposal facility*). Detailed surface and subsurface investigations and *activities* at a site to determine the radiological conditions at the site or to evaluate candidate *disposal* sites to obtain information to determine the suitability of the site for a *disposal facility* and to evaluate the long term performance of a *disposal facility* at the site.

ⓘ *Site characterization* is a stage in the *siting* of a *disposal facility*; it follows *area survey* and precedes *site confirmation* for a *disposal facility*.

ⓘ *Site characterization* may also refer to the *siting process* for any other *authorized facility*.

See also *site evaluation*, which includes *site characterization* and is not specific to a *disposal facility* site, and *area survey*.

characterization of waste

See *characterization* (2).

chemisorption

See *sorption*.

child

ⓘ In dosimetry (e.g. in tables of *dose per unit intake* values), a *child* is often assumed to be a 10 year old. If such an assumption is made, it should be clearly stated.

See also *infant* and *reference individual*.

chronic intake

See *intake* (2).

cladding

1.   An external layer of material applied directly to another material to provide protection in chemically reactive conditions (e.g. *cladding* over ferritic material to prevent corrosion).

2.   Typically, the tube of material that houses *nuclear fuel* pellets and provides the *containment* (means of *confinement*) of radionuclides produced during fission.

ⓘ *Cladding* may also provide structural support.

ⓘ The *cladding* tube, together with the end cups or plugs, also typically provides structural support.

cleanup

See *decommissioning* (1).

clearance

1.   Removal of *regulatory control* by the *regulatory body* from *radioactive material* or *radioactive* objects within notified or authorized *facilities and activities*.

ⓘ Removal from *regulatory control* in this context refers to *regulatory control* applied for *radiation protection* purposes.

ⓘ Conceptually, *clearance* — freeing certain materials or objects in authorized *facilities and activities* from further *control* — is closely linked to, but distinct from and not to be confused with, *exemption* — determining that *controls* do not need to be applied to certain *sources* and *facilities and activities*.

ⓘ Various terms (e.g. 'free *release*') are used in different States to describe this concept.

ⓘ A number of issues relating to the concept of *clearance* and its relationship to other concepts were resolved in RS-G-1.7 [20].

2.   The net effect of the biological *processes* by which radionuclides are removed from a tissue, organ or area of the body.

ⓘ The ***clearance rate*** is the rate at which these biological *processes* occur.

clearance level

See *level*.

clearance rate

See *clearance* (2).

cliff edge effect

An instance of severely abnormal conditions caused by an abrupt transition from one status of a facility to another following a small *deviation* in a parameter or a small variation in an input value.

ⓘ In a nuclear power plant or *nuclear fuel cycle facility*, a *cliff edge effect* is an instance of severely abnormal facility behaviour caused by an abrupt transition from one facility status to another following a small *deviation* in a facility parameter, and thus a sudden large variation in facility conditions in response to a small variation in an input.

closed nuclear fuel cycle

See *nuclear fuel cycle*.

closure

1.   Administrative and technical actions directed at a *disposal facility* at the end of its *operating lifetime* — for example, covering of the disposed *waste* (for a *near surface disposal facility*) or backfilling and/or sealing (for a *geological disposal facility* and the passages leading to it) — and the termination and completion of *activities* in any associated structures.

ⓘ For other types of *facilities*, the term *decommissioning* is used.

ⓘ The terms *siting, design, construction, commissioning, operation* and *decommissioning* are normally used to delineate the six major stages of the *lifetime* of an *authorized facility* and of the associated *licensing process*. In the special case of *disposal facilities* for *radioactive waste, decommissioning* is replaced in this sequence by *closure*.

2.   [The completion of all operations at some time after the emplacement of *spent fuel* or *radioactive waste* in a *disposal facility*. This includes the final engineering or other work required to bring the *facility* to a condition that will be safe in the long term.] (See Ref. [11].)

cloud shine

Gamma *radiation* from radionuclides in an airborne plume.

See also *ground shine*.

> ***sky shine.*** Radiation emitted upwards and deflected by the air back down to the ground.
>
> ⓘ The presence of *sky shine* could result in an increase in neutron flux rates further away from the facility.
>
> ⓘ *Sky shine* can be an important consideration in health physics for high energy experimental accelerator facilities as well as installations with medical linear accelerators for radiation therapy, in relation to the evaluation of shielding designs and to environmental monitoring.

coincidence (as a feature of design)

A feature of *protection system design* such that two or more overlapping or simultaneous output signals from several *channels* are necessary in order to produce a *protective action* signal by the *logic*.

collective dose

See *dose concepts*.

combustible material

A material in solid, liquid or gaseous state capable of igniting, burning, supporting combustion or releasing flammable vapour when subject to specific conditions such as fire or heat.

commissioning

The *process* by means of which *systems* and *components* of *facilities and activities*, having been constructed, are made operational and verified to be in accordance with the *design* and to have met the required performance criteria.

> ⓘ *Commissioning* may include both non-nuclear and/or non-*radioactive* and nuclear and/or *radioactive* testing.
>
> ⓘ The terms *siting, design, construction, commissioning, operation* and *decommissioning* are normally used to delineate the six major stages of the *lifetime* of an *authorized facility* and of the associated *licensing process*. In the special case of *disposal facilities* for *radioactive waste, decommissioning* is replaced in this sequence by *closure*.

committed dose

1.   See *dose concepts*.

2.   See *dose* (2).

committed effective dose

See *dose quantities*.

committed equivalent dose

See *dose quantities*.

36

common cause failure

See *failure*.

common mode failure

See *failure*.

competent authority

1.   Any body or authority designated or otherwise recognized as such for any purpose in connection with [the Transport] Regulations. (See SSR-6 (Rev. 1) [2].)

> ⓘ This term is used in the Transport Regulations [2] for consistency with terminology used in the wider field of regulation of the transport of dangerous goods.

> ⓘ This term is also used for organizations that have responsibilities for issuing and receiving notification and information under the Convention on Early Notification of a Nuclear Accident or are authorized to make and receive requests for and to accept offers of assistance under the Convention on Assistance in the Case of a Nuclear Accident or Radiological Emergency.

> ⓘ Otherwise, the more general term *regulatory body* should be used in the context of safety, with which *competent authority* is essentially synonymous.

2.   A governmental organization or institution that has been designated by a State to carry out one or more *nuclear security* functions.

> ⓘ For example, *competent authorities* may include *regulatory bodies,* law enforcement, customs and border control, intelligence and security agencies or health agencies, etc.

compliance assurance

A systematic programme of measures applied by a *regulatory body* that is aimed at ensuring that the provisions of regulations are met in practice.

> ⓘ *Compliance assurance* is a systematic programme of measures applied by a *competent authority* that is aimed at ensuring that the provisions of [the Transport] Regulations are met in practice. (See SSR-6 (Rev. 1) [2].)

> ⓘ The term may be used in a variety of contexts with essentially the same meaning, but often without explicit definition.

component

See *structures, systems and components*.

compromise

The accidental or deliberate violation of *confidentiality*, loss of *integrity*, or loss of *availability* of an *information object*.

> ⓘ The verb is also used more generally in various Nuclear Security Series publications, without specific definition, to describe security (or some other desirable characteristic) being degraded in some way.

computational model

See *model*.

computer-based systems

Technologies that create, provide access to, process, compute, communicate or store digital information, or perform, provide or control services involving such information.

    ⓘ These technologies may be physical or virtual. They may include but are not limited to: desktop, laptop, tablet and other personal computers; smart phones; mainframe computers; servers; virtual computers; software applications; databases; removable media; digital instrumentation and control devices; programmable *logic* controllers; printers; network devices; and embedded components and devices.

computer security

A particular aspect of *information security* that is concerned with the protection of *computer-based systems* against *compromise*.

    ⓘ Synonymous with **cyber-security** and **IT security**. *Computer security* is the preferred term in IAEA publications.

computer security incident

An occurrence that actually or potentially jeopardizes the confidentiality, integrity or availability of a computer-based system (including information), or that constitutes a violation or imminent risk of violation of security policies.

computer security level

The strength of protection required to meet *computer security* requirements for a function related to *nuclear security, safety, nuclear material accounting and control* and/or *sensitive information* management.

computer security measures

Measures intended to prevent, detect or delay, respond to, and mitigate the consequences of *malicious act*s or other acts that could compromise computer security.

computer security programme

A plan for the implementation of the computer security strategy specifying organizational roles, responsibilities and procedures. The programme specifies and details the means for achieving the computer security goals and is a part of (or linked to) the overall security plan.

computer security zone

A group of systems having common physical and/or logical boundaries — and, if necessary, arranged using additional criteria — that is assigned a common computer security level to simplify the administration, communication and application of computer security measures.

computerized procedures system

A system that presents plant procedures in a computer based rather than a paper based format.

concept of operations

1.   A brief description of an ideal response to a postulated *nuclear or radiological emergency*, used to ensure that all those personnel and organizations involved in the development of a capability for *emergency response* share a common understanding.

2.   The proposed design in terms of how it will be operated to perform its functions, which includes the various roles of personnel and how they will be organized, managed and supported.

  ⓘ The *concept of operations* describes how the plant is operated ('operating philosophy') and includes aspects such as the number and composition of *operating personnel* and how they operate the plant under normal and abnormal conditions.

conceptual model

See *model*.

condition based maintenance

See *maintenance*: *predictive maintenance*.

condition indicator

See *indicator*.

condition monitoring

See *monitoring* (2).

conditional probability value (CPV)

The upper bound for the conditional probability that a particular type of *event* will cause unacceptable radiological consequences.

  ⓘ The term is used in the detailed *event screening process* for *site evaluation*.

conditional risk

See *risk* (3).

conditioning

See *radioactive waste management* (1).

confidentiality

The property that information is not made available or disclosed to unauthorized individuals, entities or processes.

configuration baseline

A set of configuration items formally designated and fixed at a specific time during an item's *life* cycle.

configuration management

The *process* of identifying and documenting the characteristics of a *facility's structures, systems and components* (including computer *systems* and software), and of ensuring that changes to these characteristics are properly developed, assessed, approved, issued, implemented, verified, recorded and incorporated into the *facility* documentation.

    ⓘ 'Configuration' is used in the sense of the physical, functional and operational characteristics of the *structures, systems and components* and parts of a *facility*.

confinement

Prevention or *control* of *releases* of *radioactive material* to the *environment* in *operation* or in *accidents*.

    ⓘ *Confinement* is closely related in meaning to *containment*, but *confinement* is typically used to refer to the *safety function* of preventing the 'escape' of *radioactive material*, whereas *containment* refers to the means for achieving that function.

    ! The Transport Regulations [2] adopt a different distinction between *confinement* and *containment*, namely that *confinement* relates to preventing *criticality* and *containment* to preventing *releases* of *radioactive material* (see *confinement system* and *containment system*).

    ⓘ The main issue here is the differences in usage between the *safety* of *nuclear installations* and *safety* in the *transport* of *radioactive material*. Both terms, *containment* and *confinement*, are used in both areas (in the Transport Regulations [2], in the form of *confinement system* and *containment system*), and the usages of *containment* are (it seems) conceptually consistent, but the usages of *confinement* are not. *Confinement* in *nuclear safety* is the *safety function* that is performed by the *containment*.

confinement system

The assembly of *fissile material* and *packaging components* specified by the designer and agreed to by the *competent authority* as intended to preserve *criticality safety*. (See SSR 6 (Rev. 1) [2].)

    ! This usage is specific to the Transport Regulations [2].

See *confinement* for more general usage.

    ⓘ A *confinement system* as defined in the Transport Regulations [2] has the primary function of controlling *criticality* (as compared with the *containment system*, the function of which is to prevent leakage of *radioactive material*).

    ⓘ Discussions with experts in the field confirmed that a distinct term is needed to describe this distinct concept, and that *confinement* is the term that has become established, but failed to reveal any compelling reasons for the choice of that particular word.

consequence assessment

See *assessment* (1).

consignee

Any person, organization or government that is entitled to take delivery of a *consignment*. (See SSR-6 (Rev. 1) [2].)

    ⓘ In some publications in the Nuclear Security Series, the term **receiver** has also been used for this concept.

consignment

Any *package* or *packages*, or load of *radioactive material*, presented by a *consignor* for *transport*. (See SSR-6 (Rev. 1) [2].)

consignor

Any person, organization or government that prepares a *consignment* for *transport*. (See SSR-6 (Rev. 1) [2].)

    ⓘ In some publications in the Nuclear Security Series, the term ***shipper*** has also been used for this concept, defined as "Any person, organization or government that prepares or offers a consignment of *radioactive material* for *transport* (i.e. the consignor)" [6, 7]. The term *shipper* is used in safeguards, and therefore relates particularly to the preparation and shipment of nuclear material. The only substantive difference between the terms *shipper* and *consignor* appears to be that movements of nuclear material from one *material balance area* to another within a facility (transfers) have a *shipper*, whereas the term *consignor* necessarily implies transport of radioactive material in the public domain.

constraint

A prospective and *source* related value of *individual dose* (see *dose constraint*) or of individual *risk* (see *risk constraint*) that is used in *planned exposure situations* as a parameter for the *optimization of protection and safety* for the *source*, and that serves as a boundary in defining the range of options in *optimization*.

construction

The *process* of manufacturing and assembling the *components* of a *facility*, the carrying out of civil works, the installation of *components* and equipment, and the performance of associated tests.

    ⓘ The terms *siting, design, construction, commissioning, operation* and *decommissioning* are normally used to delineate the six major stages of the *lifetime* of an *authorized facility* and of the associated *licensing process*. In the special case of *disposal facilities* for *radioactive waste*, *decommissioning* is replaced in this sequence by *closure*.

consumer product

A device or manufactured item into which radionuclides have deliberately been incorporated or produced by *activation*, or which generates *ionizing radiation*, and which can be sold or made available to *members of the public* without special surveillance or *regulatory control* after sale.

    ⓘ *Consumer products* include items such as smoke detectors and luminous dials into which radionuclides have deliberately been incorporated and ion generating tubes. It does not include building materials, ceramic tiles, spa waters, minerals and foodstuffs, and it excludes products and appliances installed in public places (e.g. tritium exit signs).

container, waste

See *waste container*.

containment

1.   Methods or physical structures designed to prevent or *control* the *release* and the *dispersion* of *radioactive substances*.

   ⓘ Although related to *confinement*, *containment* is usually used to refer to methods or structures that perform a *confinement* function in *facilities and activities*, namely preventing or controlling the *release* of *radioactive substances* and their *dispersion* in the *environment*.

   See *confinement* for a more extensive discussion.

   ⓘ In the context of *waste disposal*, the *containment* of the radionuclides associated with the *waste* is through the provision of engineered *barriers* and natural *barriers* — including the *waste form* and *packaging*, *backfill* materials, the host *environment* and geological formations — for *confinement* of the radionuclides within the waste matrix, the *packaging* and the *disposal facility* and thus its *isolation* from the *environment*.

2.   Structural elements (cans, gloveboxes, storage cabinets, rooms, vaults, etc.), which are used to establish the physical integrity of an area or items and to maintain the continuity of knowledge of *nuclear material*.

containment system

1.   A structurally closed physical *barrier* (especially in a *nuclear installation*) designed to prevent or *control* the *release* and the *dispersion* of *radioactive substances*, and its associated systems.

2.   The assembly of *components* of the *packaging* specified by the designer as intended to retain the *radioactive material* during *transport*. (See SSR-6 (Rev. 1) [2].)

   ⓘ *Containment system* is consistent with the general *safety* usage of *containment* (definition 1), unlike *confinement system* and *confinement*.

contamination

1.   *Radioactive substances* on surfaces, or within solids, liquids or gases (including the human body), where their presence is unintended or undesirable, or the *process* giving rise to their presence in such places.

   ⓘ Also used less formally to refer to a quantity, namely the *activity* on a surface (or on a unit area of a surface).

   ⓘ *Contamination* does not include residual material remaining at a site after the completion of *decommissioning*.

   ! The term *contamination* may have a connotation that is not intended. The term *contamination* refers only to the presence of *radioactivity*, and gives no indication of the magnitude of the *hazard* involved.

   ! The term 'contaminated evidence' connotes a different meaning for the forensic scientist and for the *nuclear forensic* scientist. In general forensic science usage, 'contaminated evidence' refers to evidence contaminated with extraneous material, and thus compromised. Care should be taken to ensure that there is no possibility of confusion in referring to the presence of radionuclides on or within physical evidence: the term 'evidence contaminated with radionuclides' should be used to clarify this meaning.

2.   The presence of a *radioactive substance* on a surface in quantities in excess of 0.4 Bq/cm^2 for beta and gamma emitters and *low toxicity alpha emitters*, or 0.04 Bq/cm^2 for all other alpha emitters. (See SSR-6 (Rev. 1) [2].)

   ⓘ This is a regulatory definition of *contamination*, specific to the Transport Regulations [2]. Levels below 0.4 Bq/cm^2 or 0.04 Bq/cm^2 would still be considered *contamination* according to the scientific definition (1).

fixed contamination. *Contamination* other than *non-fixed contamination.* (See SSR-6 (Rev. 1) [2].)

non-fixed contamination. *Contamination* that can be removed from a surface during routine conditions of *transport.* (See SSR-6 (Rev. 1) [2].)

ⓘ Also termed ***removable contamination***.

contamination zone

A zone in which special *protective actions* are necessary, owing to actual or potential air *contamination* or loose surface *contamination* in excess of a specified level.

contingency plan

Predefined sets of actions for response to unauthorized acts indicative of attempted *unauthorized removal* or *sabotage*, including threats thereof, designed to effectively counter such acts.

contributory hazard

See *hazard*.

control

1.   The function or power or (usually as *controls*) means of directing, regulating or restraining.

ⓘ It should be noted that the usual meaning of the English word *control* in *safety* related contexts is somewhat 'stronger' (more active) than that of its usual translations and other similar words in some other languages. For example, *control* typically implies not only checking or *monitoring* something but also ensuring that corrective or *enforcement* measures are taken if the results of the checking or *monitoring* indicate such a need. This is in contrast, for example, to the more limited usage of the equivalent word in French and Spanish.

control (of nuclear material).

Activities, devices, systems and procedures that ensure that the continuity of knowledge (e.g. location, quantitative measurements) about *nuclear material* is maintained.

See also *system for nuclear material accounting and control*.

institutional control.

Control of a *radioactive waste* site by an authority or institution designated under the laws of a State. This *control* may be active (*monitoring, surveillance,* remedial work) or passive (land use *control*) and may be a factor in the *design* of a *facility* (e.g. a *near surface disposal facility*).

ⓘ Most commonly used to describe *controls* over a *disposal facility* after *closure* or a *facility* undergoing *decommissioning*.

ⓘ Also refers to the *controls* placed on a site that has been released from *regulatory control* under the condition of observing specified restrictions on its future use to ensure that these restrictions are complied with.

ⓘ The term *institutional control* is more general than *regulatory control* (i.e. *regulatory control* may be thought of as a special form of *institutional control*).

ⓘ *Institutional control* measures may be passive, they may be imposed for reasons not related to *protection* or *safety* (although they may nevertheless have some impact on *protection and safety*), they may be applied by organizations that do not meet the definition of a *regulatory body*, and they may apply in situations which do not fall within the scope of *facilities and activities*. As a result, some form of *institutional control* may be considered more likely to endure further into the future than *regulatory control*.

regulatory control

1. Any form of *control* or regulation applied to *facilities and activities* by a *regulatory body* for reasons relating to *nuclear safety* and *radiation protection* or to *nuclear security*.

ⓘ In several publications in the IAEA Nuclear Security Series, a broadly similar definition is established: "Any form of institutional *control* applied to *nuclear material* or *other radioactive material, associated facilities*, or *associated activities* by any *competent authority* as required by the legislative and regulatory provisions related to safety, security, or safeguards" [7, 8, 9].

ⓘ '*out of regulatory control*' refers to the absence of the direct control over material by an *authorized person* that is or would be mandated by *regulatory control* for such material, i.e. a situation in which *nuclear material* or other *radioactive material* is present without an appropriate *authorization*, either because *controls* have failed for some reason, or because they never existed. Material might therefore be designated as *out of regulatory control* even when some aspects of *regulatory control* are in place.

2. [Any form of *control* or regulation applied to *facilities* or *activities* by a *regulatory body* for reasons relating to *radiation protection* or to the *safety* or security of *radioactive sources*.] (See Ref. [21].)

! This definition is particular to the Code of Conduct on the Safety and Security of Radioactive Sources [21].

2. A standard of comparison used to check the inferences deduced from an experiment.

ⓘ In *protection and safety*, a *control* is most commonly a sample or a group of people that has not been exposed to *radiation* from a particular *source*; the occurrence of particular effects in a sample or group of people that has been exposed is compared with that in the *control* to provide some indication of the effects that may be attributable to the *exposure*.

ⓘ For example, a case–*control* study is a common type of epidemiological study in which the incidence of *health effects* (the 'cases') in a population that has been exposed to *radiation* from a particular *source* is compared with the incidence in a similar population (the '*control*') that has not been exposed, to investigate whether *exposure* due to that *source* may be causing *health effects*.

control point

The depth at which the seismic ground motion response spectrum is defined by the seismic hazard assessment.

ⓘ Typical *control point* locations are at free field ground surface, at the outcrop of bedrock or at any other specified depth in the soil profile.

controlled area

See *area*.

controlled state

See *plant states (considered in design)*.

conveyance

(a)   For *transport* by road or rail: any *vehicle*;

(b)   For *transport* by water: any *vessel*, or any hold, compartment, or *defined deck area* of a *vessel*;

(c)   For *transport* by air: any *aircraft*. (See SSR-6 (Rev. 1) [2].)

core components

The elements of a reactor core, other than *fuel assemblies*, that are used to provide structural support of the core construction, or the tools, devices or other items that are inserted into the reactor core for core *monitoring*, flow *control* or other technological purposes and are treated as core elements.

> ⓘ Examples of *core components* are *reactivity control* devices or *shutdown* devices, neutron *sources*, dummy *fuel*, *fuel* channels, instrumentation, flow restrictors and *burnable absorbers*.

corrective maintenance

See *maintenance*.

cost–benefit analysis

See *analysis*.

countermeasure

An action aimed at alleviating the radiological consequences of an *accident*.

> ⓘ *Countermeasures* are forms of *intervention*. They may be *protective actions* or *remedial actions*, and these more specific terms should be used where possible. The terms *countermeasure* and *agricultural countermeasure* are not used in GSR Part 7 [22].

> **agricultural countermeasure.** Action taken to reduce *contamination* of *food*, or agricultural or forestry products before they reach consumers.

> ⓘ Note that restrictions on the sale, movement or use of contaminated *food*, or agricultural or forestry products (i.e. measures to prevent their reaching consumers) are *countermeasures*, but are not considered to be *agricultural countermeasures*.

'cradle to grave' approach

An approach in which all the stages in the *lifetime* of a *facility*, *activity* or product are taken into consideration.

> ⓘ For example, the *'cradle to grave' approach* to the *safety* and *security* of *radioactive sources*.

See *ageing management*.

See *life cycle management*.

crime scene

A site containing records of activities, alleged to be a crime.

> **crime scene operations.** The procedures that aim to control access at a *crime scene*, to document the scene as it was first encountered, and to recognize, collect, package and remove from the scene all relevant evidence.

> **radiological crime scene.** A *crime scene* at which a criminal act or intentional unauthorized act involving *nuclear or other radioactive material* has taken place or is suspected.

criminal act

See *malicious act*.

critical (adjective)

> ! In view of the number of special meanings attached to this word, particular care should be taken when using the adjective 'critical' in its more common English senses (i.e. to mean extremely important, or as a derivative of the verb 'criticize').

1.   Having a *reactivity* of zero.

> ⓘ Also used, more loosely, when the *reactivity* is greater than zero.

See *criticality*.

2.   Relating to the highest *doses* or *risks* attributable to a specified *source*.

> ⓘ As in, for example, *critical exposure pathway* or *critical* radionuclide.

3.   Capable of sustaining a nuclear chain reaction.

> ⓘ As in, for example, *critical* mass.

critical assembly

An assembly containing *fissile material* intended to sustain a controlled fission chain reaction at a low power level, used to investigate reactor core geometry and composition.

> ⓘ A *critical assembly* — as a device that is designed and used to sustain nuclear reactions — may be subject to frequent changes to the configuration of the reactor core and the lattice, and may frequently be used as a mock-up of a configuration of a reactor core.

[critical group]

A group of *members of the public* which is reasonably homogeneous with respect to its *exposure* for a given *radiation source* and is typical of individuals receiving the highest *effective dose* or *equivalent dose* (as applicable) from the given *source*.

See *representative person*.

[*hypothetical critical group*]. A hypothetical group of individuals which is reasonably homogeneous with respect to the *risk* to which its members are subject from a given *radiation source*, and is representative of the individuals likely to be most at *risk* from the given *source*.

critical level

See *minimum significant activity (MSA)*.

criticality

The state of a nuclear chain reacting medium when the chain reaction is just self-sustaining (or *critical*), i.e. when the *reactivity* is zero.

ⓘ Often used, slightly more loosely, to refer to states in which the *reactivity* is greater than zero.

criticality accident

See *accident* (1).

criticality safety index (CSI)

A number assigned to a *package*, *overpack* or *freight container* containing *fissile material* that is used to provide *control* over the accumulation of *packages*, *overpacks* or *freight containers* containing *fissile material*. (See SSR-6 (Rev. 1) [2].)

ⓘ The *procedure* for calculating the *criticality safety index* and the restrictions on the total sum of the *criticality safety index* in a *freight container* or aboard a *conveyance* are prescribed in sections V and VI of the Transport Regulations [2].

crust, Earth's

See *Earth's crust*.

[curie (Ci)]

Unit of *activity*, equal to 3.7×10^{10} Bq (exactly).

ⓘ Superseded by the *becquerel (Bq)*. *Activity* values may be given in Ci (with the equivalent in Bq in parentheses) if they are being quoted from a reference which uses that unit.

ⓘ Originally, the *activity* of a gram of radium.

cyber-attack

A *malicious act* with the intention of stealing, altering, preventing access to or destroying a specified *target* through unauthorized access to (or actions within) a susceptible *computer-based system*.

ⓘ Such an act is characterized as a *cyber-attack* because it is directed at or exploits a *computer-based system*. The means by which that system is attacked may be electronic or physical.

cyber-security

See *computer security*.

D

dangerous source

See *source* (2).

[de minimis]

> ! The appropriate terminology of *exemption*, *clearance*, etc., should be used in *IAEA publications*.

> ⓘ A general term used historically to describe concepts that would now be referred to by terms such as *exemption* or *clearance*. The term is also sometimes used to describe a related (and controversial) philosophy that *assessments* of *collective dose* should exclude that portion delivered at very low *individual dose* rates.

> ⓘ The term *de minimis* is still used in some specific contexts, such as the London Convention 1972 [23].

> ⓘ Derived from the Latin maxim '*de minimis* non curat lex' (the law does not concern itself with trivia).

decay constant, λ

For a radionuclide in a particular energy state, the quotient of dP by dt, where dP is the likelihood of a single nucleus undergoing a spontaneous nuclear transition from that energy state in the time interval dt.

$$\lambda = \frac{dP}{dt} = -\frac{1}{N}\frac{dN}{dt} = \frac{A}{N}$$

where N is the number of nuclei of concern existing at time t and A is the *activity*.

> ⓘ The decay constant is a constant of proportionality describing the likelihood that a single nucleus will undergo a spontaneous nuclear transition from a higher energy state to a lower energy state within a differential time period. It also corresponds to:

$$\lambda = -\lim_{\Delta t \to 0} \frac{\Delta N / N}{\Delta t} = -\frac{1}{N}\frac{dN}{dt} = \frac{A}{N}$$

> ⓘ Unit: reciprocal second (s^{-1}).

> ⓘ The *activity* is the *decay constant* multiplied by the number of nuclei of the radionuclide present.

> ⓘ The *decay constant* is related to the *radioactive half-life*, $T_{1/2}$, of the radionuclide by the expression:

$$\lambda = \frac{\ln 2}{T_{1/2}}$$

decision limit

See *minimum significant activity (MSA)*.

decommissioning

1.   Administrative and technical actions taken to allow the removal of some or all of the *regulatory controls* from a *facility*.

- ! This does not apply for that part of a *disposal facility* in which *radioactive waste* is emplaced, or for certain *facilities* used for the *disposal* of *naturally occurring radioactive material (NORM)* or of residues from the mining and processing of *radioactive* ores. For all of these the term *closure* is used instead of *decommissioning*.

- ! *Decommissioning* typically includes *dismantling* of the *facility* (or part thereof) to reduce the associated *radiation risks*, but in the IAEA's usage this need not be the case. A facility could, for example, be *decommissioned* without *dismantling* and the existing structures subsequently put to another use (after *decontamination*).

- ⓘ The use of the term *decommissioning* implies that no further use of the *facility* (or part thereof) for its existing purpose is foreseen.

- ⓘ Actions for *decommissioning* are taken at the end of the *operating lifetime* of a *facility* to retire it from service with due regard for the health and *safety* of workers and *members of the public* and *protection of the environment*.

- ⓘ Subject to national legal and regulatory *requirements*, a *facility* (or its remaining parts) may also be considered decommissioned if it is incorporated into a new or existing *facility*, even if the site on which it is located is still under *regulatory control* or *institutional control*.

- ⓘ The actions will need to be such as to ensure the long term *protection* of the public and *protection of the environment*, and typically include reducing the levels of residual radionuclides in the materials and on the site of the *facility* so that the materials can be safely recycled, reused or disposed of as *exempt waste* or as *radioactive waste* and the site can be released for *unrestricted use* or otherwise reused.

- ⓘ *Decommissioning* can entail activities that are similar to *remediation* (also an authorized process), such as removal of contaminated soil from an area within the authorized boundary of a *facility*, but in this case, such removals are normally referred to as **cleanup** activities and are typically performed under the *authorization* for *decommissioning*.

- ⓘ The terms *siting, design, construction, commissioning, operation* and *decommissioning* are normally used to delineate the six major stages of the *lifetime* of an *authorized facility* and of the associated *licensing process*. In the special case of *disposal facilities* for *radioactive waste*, *decommissioning* is replaced in this sequence by *closure*.

decommissioning plan. A document containing detailed information on the proposed *decommissioning* of a *facility*.

- ⓘ The approved *decommissioning plan* describes the *actions* (including *decontamination* and/or the removal of *structures, systems and components*) to be taken in carrying out *procedures, processes* and work *activities* for the purposes of *decommissioning*.

- ⓘ The *decommissioning plan* is considered to have been fulfilled when the approved *end state* of the *facility* has been reached.

dismantling. The taking apart, disassembling and tearing down of the *structures, systems and components* of a *facility* for the purposes of *decommissioning*.

- ⓘ The two main types of *dismantling* are ***immediate dismantling*** and ***deferred dismantling***.

    Deferred dismantling is deferred after *permanent shutdown*. For a *nuclear installation*, the *nuclear fuel* is first removed. Part or all of a *facility* containing *radioactive material* is either processed or put in such a condition that it can be put into *storage*. The *facility* is maintained until it can subsequently be *decontaminated* and/or *dismantled*.

ⓘ *Deferred dismantling* can involve the early *dismantling* of some parts of the *facility* and the early processing of some *radioactive material* and its removal from the *facility*, as preparatory steps for *storage* of the remaining parts of the *facility*.

Immediate dismantling begins shortly after *permanent shutdown*. Equipment and the *structures, systems and components* of a *facility* containing *radioactive material* are removed and/or are *decontaminated* to a level that permits the removal of *regulatory control* from the *facility* and its *release*, either for *unrestricted use* or with restrictions on its future use.

entombment. The encasing of part or all of a *facility* in a structure of long lived material for the purposes of *decommissioning*.

ⓘ *Entombment* is not considered an acceptable strategy for *decommissioning* a *facility* following planned *permanent shutdown*.

ⓘ *Entombment* may be considered acceptable only under exceptional circumstances (e.g. following a *severe accident*). In this case, the *entombment* structure is maintained and *surveillance* is continued until the *radioactive* inventory decays to a level permitting termination of the *licence* and unrestricted *release* of the structure.

2.   [All steps leading to the release of a *nuclear facility*, other than a *disposal facility*, from *regulatory control*. These steps include the *processes* of *decontamination* and *dismantling*.] (See Ref. [11].)

decommissioning plan

See *decommissioning* (1).

decontamination

The complete or partial removal of *contamination* by a deliberate physical, chemical or biological *process*.

ⓘ This definition is intended to include a wide range of *processes* for removing *contamination* from people, equipment and buildings, but to exclude the removal of radionuclides from within the human body or the removal of radionuclides by natural weathering or *migration processes*, which are not considered to be *decontamination*.

See also *remediation*.

decontamination factor

The ratio of the *activity* per unit area (or per unit mass or volume) before a particular *decontamination* technique is applied to the *activity* per unit area (or per unit mass or volume) after application of the technique.

ⓘ This ratio may be specified for a particular radionuclide or for gross *activity*.

ⓘ The *background activity* may be deducted from the *activity* per unit area both before and after a particular *decontamination* technique is applied.

decorporation

The action of the biological processes by means of which incorporated radionuclides are removed from the human body.

ⓘ *Decorporation* may be promoted by chemical or biological agents.

deep sea disposal

See *disposal* (3).

defence in depth

1.   A hierarchical deployment of different levels of diverse equipment and *procedures* to prevent the escalation of *anticipated operational occurrences* and to maintain the effectiveness of physical *barriers* placed between a *radiation source* or *radioactive material* and *workers, members of the public* or the *environment*, in *operational states* and, for some *barriers*, in *accident conditions*.

ⓘ The objectives of *defence in depth* are:

(a) To compensate for human induced *events* and *component failures*;

(b) To maintain the effectiveness of the *barriers* by averting damage to the *facility* and to the *barriers* themselves;

(c) To protect *workers, members of the public* and the *environment* from harm in *accident conditions* in the *event* that these *barriers* are not fully effective.

ⓘ The Fundamental Safety Principles (IAEA Safety Fundamentals) [24] (para. 3.31) states that: "*Defence in depth* is implemented primarily through the combination of a number of consecutive and independent levels of protection that would have to fail before harmful effects could be caused to people or to the *environment*. If one level of protection or *barrier* were to fail, the subsequent level or *barrier* would be available. When properly implemented, *defence in depth* ensures that no single technical, human or organizational *failure* could lead to harmful effects, and that the combinations of *failures* that could give rise to significant harmful effects are of very low probability. The independent effectiveness of the different levels of defence is a necessary element of *defence in depth*."

ⓘ Five levels of *defence in depth* are discussed in SSR-2/1 (Rev. 1) [25] (see SSR-2/1 (Rev. 1) [25] for further information):

(a) The purpose of the first level of defence is to prevent deviations from *normal operation* and the *failure* of *items important to safety*.

(b) The purpose of the second level of defence is to detect and *control* deviations from *normal operation* in order to prevent *anticipated operational occurrences* from escalating to *accident conditions*.

(c) The purpose of the third level of defence is to prevent damage to the reactor core and *releases* of *radioactive material* requiring *off-site protective actions* and to return the plant to a *safe state* by means of inherent and/or engineered safety features, *safety systems* and procedures.

(d) The purpose of the fourth level of defence is to prevent the progress of, and to mitigate the consequences of, *accidents* that result from failure of the third level of defence by preventing accident sequences that lead to *large release of radioactive material* or *early release of radioactive material* from occurring.

(e) The purpose of the fifth and final level of defence is to mitigate radiological consequences of a *large release of radioactive material* or an *early release of radioactive material* that could potentially result from an accident.

ⓘ The International Nuclear Safety Group (INSAG) defined five levels of *defence in depth* (see Ref. [26] for further information):

(a) Level 1: Prevention of *abnormal operation* and *failures*.

(b) Level 2: *Control* of *abnormal operation* and detection of *failures*.

(c) Level 3: *Control* of *accidents* within the *design basis*.

51

(d) Level 4: *Control* of severe plant conditions, including prevention of *accident* progression and mitigation of the consequences of *severe accidents.*

(e) Level 5: Mitigation of radiological consequences of significant *releases* of *radioactive material.*

ⓘ The levels of defence are sometimes grouped into three *safety layers*: hardware, software and management *control.*

ⓘ In the context of *waste disposal*, the term *multiple barriers* is used to describe a similar concept.

2.   The combination of multiple layers of systems and measures that have to be overcome or circumvented before *nuclear security* is compromised.

ⓘ In some Nuclear Security Series publications, this term has also been defined as "The combination of successive [or multiple] layers of systems and measures for the protection of *targets* from *nuclear security threats*" [8, 9].

deferred dismantling

See *decommissioning* (1).

defined deck area

The area of the weather deck of a *vessel*, or of a *vehicle* deck of a roll-on/roll-off ship or ferry, that is allocated for the stowage of *radioactive material*. (See SSR-6 (Rev. 1) [2].)

dependability

A general term describing the overall trustworthiness of a *system*; that is, the extent to which reliance can justifiably be placed on this *system*. *Reliability, availability* and *safety* are attributes of *dependability.*

depleted uranium

See *uranium.*

derived air concentration (DAC)

A *derived limit* on the *activity concentration* in air of a specified radionuclide, calculated such that the *reference individual*, breathing air with constant *contamination* at the *DAC* with the breathing behaviour of a reference *worker* for a working year, would receive an *intake* corresponding to the *annual limit on intake* for the radionuclide in question.

ⓘ The parameter values recommended by the International Commission on Radiological Protection for calculating *DACs* are a breathing rate of 1.2 m^3/h and a working year of 2000 h [27–29].

ⓘ The breathing behaviour of a reference *worker* as defined by the International Commission on Radiological Protection [28].

derived limit

See *limit.*

design

1.   The *process* and the result of developing a concept, detailed plans, supporting calculations and specifications for a *facility* and its parts.

> ⓘ The terms *siting, design, construction, commissioning, operation* and *decommissioning* are normally used to delineate the six major stages of the *lifetime* of an *authorized facility* and of the associated *licensing process*. In the special case of *disposal facilities* for *radioactive waste, decommissioning* is replaced in this sequence by *closure*.

2.   The description of *fissile material* excepted [in the Transport Regulations], *special form radioactive material, low dispersible radioactive material, package* or *packaging* that enables such an item to be fully identified. The description may include specifications, engineering drawings, reports demonstrating compliance with regulatory *requirements*, and other relevant documentation. (See SSR-6 (Rev. 1) [2].)

> ⓘ This is a much more restricted definition than (1), and is specific to the Transport Regulations [2].

design basis

The range of conditions and *events* taken explicitly into account in the *design* of *structures, systems and components* and equipment of a *facility*, according to established criteria, such that the *facility* can withstand them without exceeding *authorized limits*.

> ⓘ Used as a noun, with the definition above. Also often used as an adjective, applied to specific categories of conditions or *events* to mean 'included in the *design basis*'; as, for example, in *design basis accident, design basis external events* and *design basis* earthquake.

design basis accident

See *plant states (considered in design)*.

design basis external events

The *external event(s)* or combination(s) of *external events* considered in the *design basis* of all or any part of a *facility*.

design basis probability value (DBPV)

A value of the annual probability for a particular type of *event* to cause unacceptable radiological consequences. It is the ratio between the *screening probability level* and the *conditional probability value*.

> ⓘ The term is used in the detailed *event screening process* for *site evaluation*.

design basis threat (DBT)

See *threat*.

design extension conditions

See *plant states (considered in design)*.

design life

See *life, lifetime*.

designated nuclear forensic laboratory

See nuclear forensic science.

detection (of a nuclear security event)

1.   A process in a *physical protection system* that begins with sensing a potentially malicious or otherwise unauthorized act and that is completed with the assessment of the cause of the alarm.

   ⓘ This definition is for use in the context of *authorized facilities* and *authorized activities*.

2.   Awareness of criminal act(s) or unauthorized act(s) with *nuclear security* implications or measurement(s) indicating the unauthorized presence of *nuclear material*, or *other radioactive material* at an *associated facility* or an *associated activity* or a *strategic location*.

   ⓘ This definition is for use in the context of material *out of regulatory control*.

   detection instrument. A complete functional system, being a combination of hardware and software (or firmware) supported by procedures for installation, calibration, maintenance and operation, used for detecting *nuclear material* or *other radioactive material*.

   ⓘ This definition is for use in relation to detection of material *out of regulatory control*.

   detection measure. Measures intended to detect a criminal or unauthorized act with nuclear security implications.

   detection system. Integrated set of *detection measures* including capabilities and resources necessary for *detection* of a criminal act or an unauthorized act with *nuclear security* implications.

   ⓘ This definition is for use in relation to detection of material *out of regulatory control*.

detection instrument

See *detection (of a nuclear security event) (2)*.

detection limit

See *minimum detectable activity (MDA)*.

detection measure

See *detection (of a nuclear security event) (2)*.

detection system

See *detection (of a nuclear security event) (2)*.

determination level

See *minimum detectable activity (MDA)*.

deterministic analysis

Analysis using, for key parameters, single numerical values (taken to have a probability of 1), leading to a single value for the result.

ⓘ In the *safety* of *nuclear installations*, for example, this implies focusing on *accident* types to substantiate compliance with established acceptance criteria in relation to the *release* of *radioactive material* and consequences, without considering the probabilities of different *event* sequences.

ⓘ Typically used with either 'best estimate' or 'conservative' values, based on expert judgement and knowledge of the phenomena being modelled.

ⓘ Contrasting terms: *probabilistic analysis* or *stochastic analysis*.

See also *probabilistic analysis*.

deterministic behaviour

Characteristic of a *system* or *component*, such that any given input sequence that is within the specifications of the item always produces the same outputs.

deterministic effect

See *health effects (of radiation)*.

deterministic timing

Characteristic of a *system* or *component*, such that the time delay between the stimulus and response has a guaranteed maximum and minimum value.

detriment

See *radiation detriment*.

deviation

A departure from specified *requirements*.

device

actuation device. A *component* that directly controls the motive power for *actuated equipment*.

ⓘ Examples of *actuation devices* include circuit breakers and relays that *control* the distribution and use of electric power, and pilot valves controlling hydraulic or pneumatic fluids.

improvised nuclear device. A device incorporating radioactive materials designed to result in the formation of a nuclear-yield reaction. Such devices may be fabricated in a completely improvised manner or may be an improvised modification to a nuclear weapon.

ⓘ Reference [6] uses the term nuclear explosive device without an explicit definition, indicating that such a device could be produced using *nuclear material* obtained by *unauthorized removal*, i.e. improvised. The term *improvised nuclear device* is used particularly to refer to a device built or adapted by a non-State actor using material *out of regulatory control*, which may also be indicative of likely characteristics of the device, but such a device is a nuclear explosive device.

inspection imaging device. An imaging device designed specifically for imaging persons or cargo *conveyances* for the purpose of detecting concealed objects on or within the human body or within cargo or a *vehicle*.

ⓘ In some types of *inspection imaging device, ionizing radiation* is used to produce images by backscatter, transmission or both.

ⓘ Other types of *inspection imaging device* utilize imaging by means of electrical and magnetic fields, ultrasound and sonar waves, nuclear magnetic resonance, microwaves, terahertz rays, millimetre waves, infrared radiation or visible light.

radiation exposure device. A device with radioactive material designed to intentionally expose members of the public to radiation.

See also *radiation generator.*

radiological dispersal device. A device to spread radioactive material using conventional explosives or other means.

ⓘ ICSANT [12] defines a 'device' in this sense as: "Any nuclear explosive device; or any radioactive material dispersal or radiation-emitting device which may, owing to its radiological properties, cause death, serious bodily injury or substantial damage to property or to the environment."

diagnostic exposure

See *exposure categories*: *medical exposure.*

diagnostic reference level

See *level*.

diffusion

The movement of radionuclides relative to the medium in which they are distributed, under the influence of a concentration gradient.

ⓘ Usually used for the movement of airborne radionuclides (e.g. from *discharges* or resulting from an *accident*) relative to the air, and for movement of dissolved radionuclides (e.g. in groundwater or surface water, from *migration* following *waste disposal*, or in surface water from *discharges*) relative to the water.

See also *advection* (where the radionuclide does not move relative to the carrying medium, but moves with it) and *dispersion*.

digital assets

Computer-based systems (or parts thereof) that are associated with or within a State's nuclear security regime.

sensitive digital assets. See *sensitive information.*

direct cause

See *cause*.

direct disposal

See *disposal* (1).

directional dose equivalent

See *dose equivalent quantities (operational)*.

discharge

1.   Planned and controlled *release* of (usually gaseous or liquid) *radioactive substances* to the *environment*.

> ⓘ Strictly, the act or *process* of releasing the *radioactive substances*, but also used to describe the *radioactive substances* released.

> ***authorized discharge.*** *Discharge* in accordance with an *authorization*.

> ***radioactive discharges.*** *Radioactive substances* arising from *sources* within *facilities and activities* which are discharged as gases, aerosols, liquids or solids to the *environment*, generally with the purpose of dilution and *dispersion*.

2.   [A planned and controlled *release* to the *environment*, as a legitimate *practice*, within *limits* authorized by the *regulatory body*, of liquid or gaseous *radioactive material* that originates from regulated nuclear *facilities* during *normal operation*.] (See Ref. [11].)

dismantling

See *decommissioning* (1).

dispersal

The spreading of *radioactive material* in the *environment*.

> ⓘ In normal language synonymous with *dispersion*, but tends to be used in a general sense, not implying the involvement of any particular *processes* or phenomena, for example the uncontrolled spreading of material that has escaped from *confinement,* or as a result of damage to (or the destruction of) a *sealed source, special form radioactive material* or *low dispersible radioactive material*.

dispersion

The spreading of radionuclides in air (***aerodynamic dispersion***) or water (***hydrodynamic dispersion***) resulting mainly from physical *processes* affecting the velocity of different molecules in the medium.

> ⓘ Often used in a more general sense combining all *processes* (including molecular *diffusion*) that result in the spreading of a plume. The terms ***atmospheric dispersion*** and ***hydrodynamic dispersion*** are used in this more general sense for plumes in air and water, respectively.

> ⓘ In usual language synonymous with *dispersal*, but *dispersion* is mostly used more specifically as defined above, whereas *dispersal* is typically (though not universally) used as a more general expression.

See also *advection* and *diffusion*.

disposal

1.   Emplacement of *waste* in an appropriate *facility* without the intention of retrieval.

ⓘ In some States, the term *disposal* is used to include *discharges* of effluents to the *environment*.

ⓘ In some States, the term *disposal* is used administratively in such a way as to include, for example, incineration of *waste* or the transfer of *waste* between *operators*.

! In *IAEA publications*, *disposal* should be used only in accordance with the more restrictive definition given above.

! In many cases, the only element of this definition that is important is the distinction between *disposal* (with no intent to retrieve) and *storage* (with intent to retrieve). In such cases, a definition is not necessary; the distinction can be made in the form of a footnote at the first use of the term *disposal* or *storage* (e.g. "The use of the term '*disposal*' indicates that there is no intention to retrieve the *waste*. If retrieval of the *waste* at any time in the future is intended, the term '*storage*' is used.").

! The term *disposal* implies that retrieval is not intended and would require deliberate action to regain access to the waste; it does not mean that retrieval is not possible.

ⓘ For *storage* in a combined *storage* and *disposal facility*, for which a decision may be made at the time of its *closure* whether to remove the *waste* stored during the *operation* of the *storage facility* or to dispose of it by encasing it in concrete, the question of intention of retrieval may be left open until the time of *closure* of the *facility*.

ⓘ Contrasted with *storage*.

direct disposal. *Disposal of spent fuel as waste.*

geological disposal. *Disposal in a geological disposal facility.*

See also *repository*.

ⓘ The term '*intermediate depth disposal*' is sometimes used for the *disposal* of *low level waste* and *intermediate level waste*, for example in boreholes (i.e. between *near surface disposal* and *geological disposal*).

near surface disposal. *Disposal*, under an engineered cover, with or without additional engineered *barriers*, in a *near surface disposal facility*.

sub-seabed disposal. *Disposal in a geological disposal facility* in the rock underlying the seabed.

2.   [The emplacement of *spent fuel* or *radioactive waste* in an appropriate *facility* without the intention of retrieval.] (See Ref. [11].)

3.   The act or *process* of getting rid of *waste*, without the intention of retrieval.

ⓘ The terms *deep sea disposal* and *seabed disposal* do not strictly satisfy definitions (1) or (2), but are consistent with the everyday meaning of *disposal* and are used as such.

deep sea disposal. *Disposal of waste* packaged in *containers* on the deep ocean floor.

! The commonly used, but informal, term 'sea dumping' should not be used in *IAEA publications*.

ⓘ As practised until 1982 in accordance with the *requirements* of the London Convention 1972 [23].

seabed disposal. Emplacement of *waste* packaged in suitable *containers* at some depth into the sedimentary layers of the deep ocean floor.

ⓘ This may be achieved by direct emplacement, or by placing the *waste* in specially designed 'penetrators' which, when dropped into the sea, embed themselves in the sediment.

disposal facility

An engineered *facility* where *waste* is emplaced for *disposal*.

ⓘ Synonymous with *repository*.

disposal system. The *system* of properties of the site for a *disposal facility*, design of the *disposal facility*, physical *structures* and items, *procedures* for *control*, characteristics of *waste* and other elements that contribute in different ways and over different timescales to the fulfilment of *safety functions* for *disposal*.

geological disposal facility. A *facility* for *radioactive waste disposal* located underground (usually several hundred metres or more below the surface) in a stable geological formation to provide long term *isolation* of radionuclides from the *biosphere*.

near surface disposal facility. A *facility* for *radioactive waste disposal* located at or within a few tens of metres of the Earth's surface.

ⓘ The *practice* of disposal of *waste* in a *near surface disposal facility* with an engineered cover is also referred to as 'shallow land burial' of waste.

disposal system

See *disposal facility*.

disposition

Consigning of, or arrangements for the consigning of, *radioactive waste* for some specified (interim or final) destination, for example for the purpose of *processing*, *disposal* or *storage*.

disused sealed source

See *source* (2): *disused source*.

disused source

See *source* (2).

diversity

The presence of two or more independent (redundant) *systems* or *components* to perform an identified function, where the different *systems* or *components* have different attributes so as to reduce the possibility of *common cause failure*, including *common mode failure*.

ⓘ Examples of such attributes are: different *operating conditions*, different working principles or different *design* teams (which provide *functional diversity*), and different sizes of equipment, different manufacturers, and types of equipment (which provide diversity of equipment) that use different physical methods (which provide ***physical diversity***).

ⓘ When the term *diversity* is used with an additional attribute, the term diversity indicates the general meaning 'existence of two or more different ways or means of achieving a specified objective', while the attribute indicates the characteristics of the different ways applied, e.g. functional diversity, equipment diversity, signal diversity. ***functional diversity.*** Application of *diversity* at the level of functions in applications in *process* engineering (e.g. for the actuation of a trip on both a pressure *limit* and a temperature *limit*).

division

The collection of items, including their interconnections, that form one *redundancy* of a redundant *system* or *safety group*.

ⓘ Divisions may include multiple channels.

dose

1.   A measure of the energy deposited by *radiation* in a target.

ⓘ For definitions of the most important such measures, see *dose quantities* and *dose concepts*.

2.   *Absorbed dose, committed equivalent dose, committed effective dose, equivalent dose, effective dose* or *organ dose*, as indicated by the context.

committed dose. *Committed equivalent dose* or *committed effective dose*.

dose and dose rate effectiveness factor (DDREF)

The ratio between the *risk* or *radiation detriment* per unit *effective dose* for high *doses* and/or *dose rates* and that for low *doses* and *dose rates*.

ⓘ Used in the estimation of *risk coefficients* for low *doses* and *dose rates* from observations and epidemiological findings at high *doses* and *dose rates*.

ⓘ Supersedes the *dose rate effectiveness factor (DREF)*.

dose assessment

See *assessment* (1).

dose coefficient

ⓘ Used by the International Commission on Radiological Protection and others as a synonym for *dose per unit intake*, but sometimes also used to describe other coefficients linking quantities or concentrations of *activity* to *doses* or *dose rates*, such as the external *dose rate* at a specified distance above a surface with a deposit of a specified *activity* per unit area of a specified radionuclide.

! To avoid confusion, the term *dose coefficient* should be used with care.

[dose commitment]

See *dose concepts*.

dose concepts

annual dose. The *dose* from *external exposure* in a year plus the *committed dose* from *intakes* of radionuclides in that year.

ⓘ The *individual dose*, unless otherwise stated.

! This is not, in general, the same as the *dose* actually delivered during the year in question, which would include *doses* from radionuclides remaining in the body from *intakes* in previous years, and would exclude *doses* delivered in future years from *intakes* of radionuclides during the year in question.

averted dose. The *dose* prevented by *protective actions*.

collective dose. The total *radiation dose* incurred by a population.

ⓘ This is the sum of all of the *individual doses* to members of the population. If the *doses* continue for longer than a year, then the annual *individual doses* must also be integrated over time.

ⓘ Unless otherwise specified, the time over which the *dose* is integrated is infinite; if a finite upper limit is applied to the time integration, the *collective dose* is described as 'truncated' at that time.

ⓘ Although the upper limit for the integral for *collective dose* could in principle be infinite, in most *assessments* of *collective dose* the component part associated with *individual doses* or *dose rates* that are higher than the thresholds for the induction of *deterministic effects* would be considered separately.

ⓘ Unless otherwise specified, the relevant *dose* is normally the *effective dose* (collective effective dose has a formal definition).

ⓘ Unit: person-sievert (person Sv). This is, strictly, just a *sievert*, but the unit person-sievert is used to distinguish the *collective dose* from the *individual dose* which a dosimeter would measure (just as, for example, 'person-hours' are used to measure the total effort devoted to a task, as opposed to the elapsed time that would be shown by a clock).

ⓘ Contrasting term: *individual dose*.

committed dose. The *lifetime dose* expected to result from an *intake*.

ⓘ This term partially supersedes *dose commitment*.

See *dose quantities*: *committed equivalent dose* and *committed effective dose*.

[*dose commitment*]. The total *dose* that would eventually result from an *event* (e.g. a *release* of *radioactive material*), a deliberate action or a finite portion of a *practice*.

ⓘ More specific and precise terms such as *committed dose* or *collective dose* should be used as appropriate.

individual dose. The *dose* incurred by an individual.

ⓘ Contrasting term: *collective dose*.

lifetime dose. The total *dose* received by an individual during his or her lifetime.

ⓘ In practice, often approximated as the sum of the *annual doses* incurred. Since *annual doses* include *committed doses*, some parts of some of the *annual doses* may not actually be delivered within the lifetime of the individual, and therefore this may overestimate the true *lifetime dose*.

ⓘ For prospective *assessments* of *lifetime dose*, a lifetime is normally interpreted as 70 years.

projected dose. The *dose* that would be expected to be received if planned *protective actions* were not taken.

residual dose. The *dose* expected to be incurred after *protective actions* have been terminated (or after a decision has been taken not to take *protective actions*).

ⓘ *Residual dose* applies for an *emergency exposure situation* or for an *existing exposure situation*.

dose constraint

A prospective and *source* related value of *individual dose* that is used in *planned exposure situations* as a parameter for the *optimization of protection and safety* for the *source*, and that serves as a boundary in defining the range of options in *optimization*.

ⓘ For *occupational exposure*, the *dose constraint* is a *constraint* on *individual dose* to *workers* established and used by *registrants* and *licensees* to set the range of options in optimizing *protection and safety* for the *source*.

ⓘ For *public exposure*, the *dose constraint* is a *source* related value established or approved by the government or the *regulatory body*, with account taken of the *doses* from planned operations of all *sources* under *control*.

ⓘ The *dose constraint* for each particular *source* is intended, among other things, to ensure that the sum of *doses* from planned *operations* for all *sources* under *control* remains within the *dose limit*.

ⓘ For *medical exposure*, the *dose constraint* is a *source* related value used in optimizing the *protection* of *carers and comforters* of *patients* undergoing *radiological procedures*, and the *protection* of volunteers subject to *exposure* as part of a programme of biomedical research.

dose conversion convention

The assumed relationship between *potential alpha energy exposure* and *effective dose*.

ⓘ Used to estimate *doses* from measured or estimated *exposure due to radon*.

See also *exposure* (4).

ⓘ Unit: mSv per $J \cdot h/m^3$.

dose equivalent

The product of the *absorbed dose* at a point in the tissue or organ and the appropriate *quality factor* for the type of *radiation* giving rise to the *dose*.

ⓘ A measure of the *dose* to a tissue or organ designed to reflect the amount of harm caused.

ⓘ For *radiation protection* purposes the quantity *dose equivalent* has been superseded by *equivalent dose*.

ⓘ *Dose equivalent* is also a term used by the International Commission on Radiation Units and Measurements in defining the following *operational quantities*: *ambient dose equivalent*, *directional dose equivalent* and *personal dose equivalent* (see *dose equivalent quantities*).

[*effective dose equivalent, H_E*]. A measure of *dose* designed to reflect the *risk* associated with the *dose*, calculated as the weighted sum of the *dose equivalents* in the different tissues of the body.

ⓘ Superseded by *effective dose*.

dose equivalent quantities (operational)

ambient dose equivalent, H*(d). The *dose equivalent* that would be produced by the corresponding aligned and expanded field in the *ICRU sphere* at a depth *d* on the radius vector opposing the direction of the aligned field.

ⓘ Parameter defined at a point in a *radiation* field. Used as a directly measurable proxy (i.e. substitute) for *effective dose* for use in *monitoring* of *external exposure*.

ⓘ The recommended value of *d* for *strongly penetrating radiation* is 10 mm.

directional dose equivalent, H'(d,Ω). The *dose equivalent* that would be produced by the corresponding expanded field in the *ICRU sphere* at a depth *d* on a radius in a specified direction Ω.

ⓘ Parameter defined at a point in a *radiation* field. Used as a directly measurable proxy (i.e. substitute) for *equivalent dose* in the skin in *monitoring* of *external exposure*.

ⓘ The recommended value of *d* for *weakly penetrating radiation* is 0.07 mm.

personal dose equivalent, H_p(d). The *dose equivalent* in soft tissue below a specified point on the body at an appropriate depth *d*.

ⓘ Parameter used as a directly measurable proxy (i.e. substitute) for *equivalent dose* in tissues or organs or (with *d* = 10 mm) for *effective dose*, in *individual monitoring* of *external exposure*.

ⓘ The recommended values of *d* are 10 mm for *strongly penetrating radiation* and 0.07 mm for *weakly penetrating radiation*.

ⓘ $H_p(0.07)$ is used for monitoring for hands and feet for all radiation types.

ⓘ $H_p(3)$ is used for monitoring exposure of the lens of the eye.

ⓘ 'Soft tissue' is commonly interpreted as the *ICRU sphere*.

ⓘ Recommended by the International Commission on Radiation Units and Measurements [30, 31] as a simplification of the two separate terms [*individual dose equivalent, penetrating, $H_p(d)$*], and [*individual dose equivalent, superficial, $H_s(d)$*], defined in Ref. [32].

dose limit

See *limit*.

dose per unit intake

The *committed effective dose* or the *committed equivalent dose* resulting from *intake*, by a specified means (usually ingestion or inhalation), of unit *activity* of a specified radionuclide in a specified chemical form.

ⓘ Values are specified in GSR Part 3 [1] and recommended by the International Commission on Radiological Protection [29].

ⓘ For *intakes*, synonymous with *dose coefficient*.

ⓘ Unit: Sv/Bq.

dose quantities

absorbed dose, D. The fundamental dosimetric quantity D, defined as:

$$D = \frac{d\bar{\varepsilon}}{dm}$$

where $d\bar{\varepsilon}$ is the mean energy imparted by *ionizing radiation* to matter in a volume element and dm is the mass of matter in the volume element.

ⓘ The energy can be averaged over any defined volume, the average *dose* being equal to the total energy imparted in the volume divided by the mass in the volume.

ⓘ *Absorbed dose* is defined at a point; for the average *dose* in a tissue or organ, see *organ dose*.

ⓘ The SI unit for *absorbed dose* is joule per kilogram (J/kg), termed the *gray* (Gy) (formerly, the *rad* was used).

> ***relative biological effectiveness (RBE) weighted absorbed dose, AD_T.*** The quantity $AD_{T,R}$, defined as:
>
> $$AD_{T,R} = D_{T,R} \times RBE_{T,R}$$
>
> where $D_{T,R}$ is the *absorbed dose* delivered by *radiation* of type R averaged over a tissue or organ T and $RBE_{T,R}$ is the *relative biological effectiveness* for *radiation* of type R in the production of *severe deterministic effects* in a tissue or organ T. When the *radiation* field is composed of different *radiation* types with different values of $RBE_{T,R}$, the *RBE weighted absorbed dose* is given by:
>
> $$AD_T = \sum_R D_{T,R} \times RBE_{T,R}$$
>
> ⓘ The unit of *RBE weighted absorbed dose* is the *gray* (Gy), equal to 1 J/kg.
>
> ⓘ *RBE weighted absorbed dose* is a measure of the *dose* to a tissue or organ, intended to reflect the *risk* of development of *severe deterministic effects*.
>
> ⓘ Values of *RBE weighted absorbed dose* to a specified tissue or organ from any type(s) of *radiation* can be compared directly.

committed effective dose, E(τ). The quantity $E(\tau)$, defined as:

$$E(\tau) = \sum_T w_T \cdot H_T(\tau)$$

where $H_T(\tau)$ is the *committed equivalent dose* to tissue or organ T over the integration time τ elapsed after an *intake* of *radioactive substances* and w_T is the *tissue weighting factor* for tissue or organ T. Where τ is not specified, it will be taken to be 50 years for *intakes* by adults and the time to the age of 70 years for *intakes* by children.

ⓘ That is, for *intakes* by children, 70 years minus the age in years: so, for example, 60 years for a 10 year old *child*.

committed equivalent dose, $H_T(\tau)$. The quantity $H_T(\tau)$, defined as:

$$H_T(\tau) = \int_{t_0}^{t_0+\tau} \dot{H}_T(t)\,\mathrm{d}t$$

where t_0 is the time of *intake*, $\dot{H}_T(t)$ is the *equivalent dose rate* at time t in tissue or organ T and τ is the integration time elapsed after an *intake* of *radioactive substances*. Where τ is not specified, it is taken to be 50 years for *intakes* by adults and the time to the age of 70 years for *intakes* by children.

ⓘ That is, for *intakes* by children, 70 years minus the age in years: so for example 60 years for a 10 year old child.

effective dose, E. The quantity *E*, defined as a summation of the tissue or organ *equivalent doses*, each multiplied by the appropriate *tissue weighting factor*:

$$E = \sum_T w_T \cdot H_T$$

where H_T is the *equivalent dose* in tissue or organ T and w_T is the *tissue weighting factor* for tissue or organ T.

From the definition of *equivalent dose*, it follows that:

$$E = \sum_T w_T \cdot \sum_R w_R \cdot D_{T,R}$$

where w_R is the *radiation weighting factor* for *radiation* type R and $D_{T,R}$ is the average *absorbed dose* in the tissue or organ T delivered by *radiation* type R.

ⓘ The SI unit for *effective dose* is joule per kilogram (J/kg), termed the *sievert* (*Sv*). An explanation of the quantity is given in annex B to Ref. [33].

ⓘ The *rem*, equal to 0.01 Sv, is sometimes used as a unit of *equivalent dose* and *effective dose*. This should not be used in *IAEA publications*, except when quoting directly from other publications, in which case the value in *sieverts* should be added in parentheses.

ⓘ *Effective dose* is a measure of *dose* designed to reflect the amount of *radiation detriment* likely to result from the *dose*.

ⓘ *Effective dose* cannot be used to quantify higher *doses* or to make decisions on the need for any medical treatment relating to *deterministic effects*.

ⓘ Values of *effective dose* from *exposure* for any type(s) of *radiation* and any mode(s) of *exposure* can be compared directly.

equivalent dose, H_T. The quantity $H_{T,R}$, defined as:

$$H_{T,R} = w_R \cdot D_{T,R}$$

where $D_{T,R}$ is the *absorbed dose* delivered by *radiation* type R averaged over a tissue or organ T and w_R is the *radiation weighting factor* for *radiation* type R.

When the *radiation* field is composed of different *radiation* types with different values of w_R, the *equivalent dose* is:

$$H_T = \sum_R w_R \cdot D_{T,R}$$

ⓘ The SI unit for *equivalent dose* is joule per kilogram (J/kg), termed the *sievert (Sv)*. An explanation of the quantity is given in annex B to Ref. [33].

ⓘ The *rem*, equal to 0.01 Sv, is sometimes used as a unit of *equivalent dose* and *effective dose*. This should not be used in *IAEA publications*, except when quoting directly from other publications, in which case the value in *sieverts* should be added in parentheses.

ⓘ *Equivalent dose* is a measure of the *dose* to a tissue or organ designed to reflect the amount of harm caused.

ⓘ *Equivalent dose* cannot be used to quantify higher *doses* or to make decisions on the need for any medical treatment relating to *deterministic effects*.

ⓘ Values of *equivalent dose* to a specified tissue or organ from any type(s) of *radiation* can be compared directly.

organ dose. The mean *absorbed dose* D_T in a specified tissue or organ T of the human body, given by:

$$D_T = \frac{1}{m_T} \int_{m_T} D.dm = \frac{\varepsilon_T}{m_T}$$

where m_T is the mass of the tissue or organ, D is the *absorbed dose* in the mass element dm and ε_T is the total energy imparted.

ⓘ Sometimes called tissue *dose*.

dose rate

1.   The *dose* per unit time.

! Although *dose rate* could, in principle, be defined over any unit of time (e.g. an *annual dose* is technically a *dose rate*), in *IAEA publications* the term *dose rate* should be used only in the context of short periods of time, for example *dose* per second or *dose* per hour.

2.   The *ambient dose equivalent* or the *directional dose equivalent*, as appropriate, per unit time, measured at the point of interest. (See SSR-6 (Rev. 1) [2].)

! This usage is specific to the Transport Regulations [2].

[dose rate effectiveness factor (DREF)]

The ratio between the *risk* per unit *effective dose* for high *dose rates* and that for low *dose rates*.

ⓘ Superseded by *dose and dose rate effectiveness factor (DDREF)*.

double contingency principle

See *single failure criterion*.

drawdown

A falling of the water level at a coastal site.

driven equipment

A *component* such as a pump or valve that is operated by a *prime mover*.

dry storage

See *storage*.

E

early effect

See *health effects (of radiation)*.

early protective actions

See *protective action* (1).

early release of radioactive material

A *release* of *radioactive material* for which *off-site protective actions* are necessary but are unlikely to be fully effective in due time.

> ⓘ See also *large release of radioactive material* and *defence in depth (1)*.

early response phase

See *emergency response phase*.

Earth's crust

The outermost solid layer of the Earth.

> ⓘ The *Earth's crust* represents less than 1% of the Earth's volume and varies in thickness from approximately 6 km beneath the oceans to approximately 60 km beneath mountain chains.

Earth's mantle

A solid layer of the Earth, approximately 2300 km thick, located between the *Earth's crust* and the Earth's core.

> ⓘ Basaltic *magma* forms from the partial melting of *mantle* rocks.

effective dose

See *dose quantities*.

[effective dose equivalent]

See *dose equivalent*.

effective half-life

See *half-life* (2).

effusive eruption

See volcanic *eruption*.

elimination, practical

See *practical elimination*.

emergency

A non-routine situation or *event* that necessitates prompt action, primarily to mitigate a *hazard* or adverse consequences for human life, health, property and the *environment*.

> ⓘ This includes *nuclear and radiological emergencies* and conventional *emergencies* such as fires, *releases* of hazardous chemicals, storms or earthquakes.

> ⓘ This includes situations for which prompt action is warranted to mitigate the effects of a perceived *hazard*.

> ⓘ Terms and definitions relating to an *emergency* are taken from GSR Part 7 [22].

See also *emergency class*.

> *nuclear or radiological emergency.* An *emergency* in which there is, or is perceived to be, a *hazard* due to:

> (a) The energy resulting from a nuclear chain reaction or from the decay of the products of a chain reaction; or

> (b) *Radiation exposure*.

> > ⓘ Points (a) and (b) approximately represent *nuclear and radiological emergencies*, respectively. However, this is not an exact distinction.

> > ⓘ *Radiation emergency* is used in some cases when an explicit distinction in the nature of the *hazard* is immaterial (e.g. national *radiation emergency* plan), and it has essentially the same meaning.

> *transnational emergency.* A *nuclear or radiological emergency* of actual, potential or perceived radiological significance for more than one State.

> ⓘ This may include:

> > (1) A *significant transboundary release* of *radioactive material* (however, a *transnational emergency* does not necessarily imply a *significant transboundary release* of *radioactive material*);

> > (2) A *general emergency* at a *facility* or other *event* that could result in a *significant transboundary release* (atmospheric or aquatic) of *radioactive material*;

> > (3) Discovery of the loss or illicit removal of a *dangerous source* that has been transported across, or is suspected of having been transported across, a national border;

> > (4) An *emergency* resulting in significant disruption to international trade or travel;

> > (5) An *emergency* warranting the taking of *protective actions* for foreign nationals or embassies in the State in which it occurs;

(6)   An *emergency* resulting or potentially resulting in *severe deterministic effects* and involving a fault and/or problem (such as in equipment or software) that could have serious implications for *safety* internationally;

(7)   An *emergency* resulting in or potentially resulting in great concern among the population of more than one State owing to the actual or perceived radiological *hazard*.

emergency action level (EAL)

See *level*.

emergency arrangements

The integrated set of infrastructural elements, put in place at the *preparedness stage*, that are necessary to provide the capability for performing a specified function or task required in response to a *nuclear or radiological emergency*.

ⓘ These elements may include: authorities and responsibilities, organization, coordination, personnel, plans, *procedures*, *facilities*, equipment or training.

emergency class

A set of conditions that warrant a similar immediate *emergency response*.

ⓘ This is the term used for communicating to the *response organizations* and to the public the level of response needed. The *events* that belong to a given *emergency class* are defined by criteria specific to the installation, *source* or *activities*, which, if exceeded, indicate classification at the prescribed level. For each *emergency class*, the initial actions of the *response organizations* are predefined.

ⓘ IAEA safety standards specify five *emergency classes*, namely *general emergency*, *site area emergency*, *facility emergency*, *alert* and *other nuclear or radiological emergency* [22]:

(a) **general emergency.** At *facilities* in *emergency preparedness category* I or II, an *emergency* that warrants taking *precautionary urgent protective actions*, *urgent protective actions* and *early protective actions* and *other response actions* on the site and off the site.

ⓘ When a *general emergency* is declared, appropriate actions are promptly taken, on the basis of the available information relating to the *emergency*, to mitigate the consequences of the *emergency* on the site and to protect people on the site and off the site.

(b) **site area emergency.** At *facilities* in *emergency preparedness category* I or II, an *emergency* that warrants taking *protective actions* and *other response actions* on the site and in the vicinity of the site.

ⓘ When a *site area emergency* is declared, actions are promptly taken: (i) to mitigate the consequences of the *emergency* on the site and to protect people on the site; (ii) to increase the readiness to take *protective actions* and *other response actions* off the site if this becomes necessary on the basis of observable conditions, reliable assessments and/or results of *monitoring*; and (iii) to conduct *off-site monitoring*, sampling and *analysis*.

(c) **facility emergency.** At *facilities in emergency preparedness category* I, II or III, an *emergency* that warrants taking *protective actions* and *other response actions* at the *facility* and on the site but does not warrant taking *protective actions* off the site.

ⓘ When a *facility emergency* is declared, actions are promptly taken to mitigate the consequences of the *emergency* and to protect people.

(d) **alert.** At facilities in *emergency preparedness category* I, II or III, an *event* that warrants taking actions to assess and to mitigate the potential consequences at the *facility*.

ⓘ When an *alert* is declared, actions are promptly taken to assess and to mitigate the potential consequences of the *event* and to increase the readiness of the on-site *response organizations*.

(e) ***other nuclear or radiological emergency.*** An *emergency* in *emergency preparedness category* IV that warrants taking *protective actions* and *other response actions* at any location.

ⓘ When such an *emergency* is declared, actions are promptly taken to mitigate the consequences of the *emergency* on the site; to protect those in the vicinity (e.g. *workers* and *emergency workers* and the public) and to determine where and for whom *protective actions* and *other response actions* are warranted.

emergency classification

The *process* whereby an authorized official classifies an *emergency* in order to declare the applicable *emergency class*.

ⓘ Upon declaration of the *emergency class*, the *response organizations* initiate the predefined *emergency response actions* for that *emergency class*.

emergency exposure situation

See *exposure situations*.

emergency phase

See *emergency response phase*.

emergency plan

A description of the objectives, policy and *concept of operations* for the response to an *emergency* and of the structure, authorities and responsibilities for a systematic, coordinated and effective response. The *emergency plan* serves as the basis for the development of other plans, *procedures* and checklists.

ⓘ *Emergency plans* are prepared at several different levels: international, national, regional, local and *facility*. They may include all *activities* planned to be carried out by all relevant organizations and authorities, or may be primarily concerned with the actions to be carried out by a particular organization.

ⓘ Details regarding the accomplishment of specific tasks outlined in an *emergency plan* are contained in *emergency procedures*.

See also *concept of operations (1)*.

emergency planning distance

The *extended planning distance* and the *ingestion and commodities planning distance*.

extended planning distance (EPD). The area around a facility within which *emergency arrangements* are made to conduct monitoring following the declaration of a *general emergency* and to identify areas warranting *emergency response actions* to be taken off the site within a period following a significant *radioactive release* that would allow the *risk* of *stochastic effects* among *members of the public* to be effectively reduced.

ⓘ The area within the *extended planning distance* serves for planning purposes and may not be the actual area in which monitoring is to be conducted to identify areas where *early protective actions* such as *relocation* are necessary.

ⓘ While efforts need to be made at the *preparedness stage* to prepare for taking effective *early protective actions* within this area, the actual area will be determined by the prevailing conditions in an emergency.

ⓘ As a precaution, some urgent actions may be warranted within the *extended planning distance* to reduce the *risk of stochastic effects* among *members of the public*.

ingestion and commodities planning distance (ICPD). The area around a facility for which *emergency arrangements* are made to take effective *emergency response actions* following the declaration of a *general emergency* in order to reduce the *risk of stochastic effects* among *members of the public* and to mitigate *non-radiological consequences* as a result of the distribution, sale and consumption of *food*, milk and drinking water and the use of commodities other than *food* that may have *contamination* from a significant *radioactive release*.

ⓘ The area within the *ingestion and commodities planning distance* serves for planning purposes to prepare for *emergency response actions* to monitor and *control* commodities, including *food*, either for domestic use or international trade.

ⓘ The actual area will be determined on the basis of the prevailing conditions in an *emergency*.

ⓘ As a precaution, some *urgent protective actions* may be warranted within the *ingestion and commodities planning distance* to prevent the ingestion of *food*, milk or drinking water and to prevent the use of commodities that may have *contamination* following a significant *radioactive release*.

emergency planning zone

The *precautionary action zone* and the *urgent protective action planning zone*.

precautionary action zone (PAZ). An area around a *facility* for which *emergency arrangements* have been made to take *urgent protective actions* in the *event* of a *nuclear or radiological emergency* to avoid or to minimize *severe deterministic effects* off the site. *Protective actions* within this area are to be taken before or shortly after a *release* of *radioactive material* or an *exposure*, on the basis of prevailing conditions at the *facility*.

urgent protective action planning zone (UPZ). An area around a *facility* for which *arrangements* have been made to take *urgent protective actions* in the event of a *nuclear or radiological emergency* to avert *doses* off the site in accordance with international *safety standards*. *Protective actions* within this area are to be taken on the basis of *environmental monitoring* — or, as appropriate, prevailing conditions at the *facility*.

emergency preparedness

The capability to take actions that will effectively mitigate the consequences of an *emergency* for human life, health, property and the *environment*.

emergency preparedness category. A category for *hazards* assessed by means of a *hazard assessment* to provide the basis for a *graded approach* to the application of the *requirements* established in GSR Part 7 [22] and for developing generically justified and optimized *arrangements* for *preparedness* and *response* for a *nuclear or radiological emergency*.

ⓘ Table 1 of GSR Part 7 [22] describes the *emergency preparedness categories*.

preparedness stage. The stage or phase at which *arrangements* for an effective *emergency response* are established prior to a nuclear or radiological emergency.

emergency preparedness category

See *emergency preparedness*.

emergency procedures

A set of instructions describing in detail the actions to be taken by *emergency workers* in an *emergency*.

emergency response

The performance of actions to mitigate the consequences of an *emergency* for human life, health, property and the *environment*.

ⓘ The *emergency response* also provides a basis for the resumption of normal social and economic activity.

emergency response action. An action to be taken in response to a *nuclear or radiological emergency* to mitigate the consequences of an *emergency* for human life, health, property and the *environment*.

ⓘ *Emergency response actions* comprise *protective actions* and *other response actions*.

ⓘ Also called *emergency action*.

other response action. An *emergency response action* other than a *protective action*.

ⓘ The most common *other response actions* are: medical *examination*, consultation and treatment; registration and long term medical follow-up; provision of psychological counselling; and public information and other actions for mitigating *non-radiological consequences* and for public reassurance.

emergency response action

See *emergency response*.

emergency response commander

The individual responsible for directing the response of all organizations responding to an *emergency* (including the response to radiological *hazards*, the response to conventional *hazards* and law enforcement).

ⓘ Also referred to as *incident commander*.

emergency response facility or location

A *facility* or location necessary for supporting an *emergency response*, for which specific functions are to be assigned at the *preparedness stage*, and which need to be usable under *emergency* conditions.

ⓘ There are two different types of *emergency response facility or location*: those established in advance (e.g. a technical support centre for a nuclear power plant) and those designated at the time of an emergency (e.g. a medical screening and triage area).

ⓘ For both types, advance preparations are necessary to ensure their operability under *emergency* conditions. Depending on the *emergency preparedness category* and on the nature of an *emergency*, an *emergency response facility* may be designated an *emergency response location*.

ⓘ For a nuclear power plant and other *facilities* in *emergency preparedness category* 1, *emergency response facilities* (which are separate from the control room and the supplementary control room) include: the technical support centre, from which technical support can be provided to the *operating personnel* in the control room in an *emergency*; the operational support centre, from which operational control can be maintained by personnel performing tasks at or near the *facility*; and the *emergency* centre, from which the on-site *emergency response* is managed.

emergency response phase

The period of time from the detection of conditions warranting an *emergency response* until the completion of all the *emergency response actions* taken in anticipation of or in response to the radiological conditions expected in the first few months of the *emergency*.

ⓘ The *emergency response phase* typically ends when the situation is under *control*, the *off-site* radiological conditions have been characterized sufficiently well to identify whether and where *food* restrictions and temporary *relocation* are required, and all required *food* restrictions and temporary *relocations* have been put into effect.

ⓘ Also called the *emergency phase*.

early response phase. The period of time, within the *emergency response phase*, from which a radiological situation is already characterized sufficiently well that a need for taking *early protective actions* and *other response actions* can be identified, until the completion of all such actions.

ⓘ The *early response phase* may last from days to weeks depending on the nature and scale of the *nuclear or radiological emergency*.

urgent response phase. The period of time, within the *emergency response phase*, from the detection of conditions warranting *emergency response actions* that must be taken promptly in order to be effective until the completion of all such actions.

ⓘ Such *emergency response actions* include *mitigatory actions* by the *operator* and *urgent protective actions* on the site and off the site.

ⓘ The *urgent response phase* may last from hours to days depending on the nature and scale of the *nuclear or radiological emergency*.

emergency services

The local *off-site response organizations* that are generally available and that perform *emergency response* functions. These may include police, firefighters and rescue brigades, ambulance services, and control teams for hazardous materials.

emergency worker

A person having specified duties as a *worker* in response to an *emergency*.

ⓘ *Emergency workers* may include *workers* employed, both directly and indirectly, by *registrants* and *licensees*, as well as personnel of *response organizations*, such as police officers, firefighters, medical personnel, and drivers and crews of vehicles used for *evacuation*.

ⓘ *Emergency workers* may or may not be designated as such in advance of an *emergency*. *Emergency workers* not designated as such in advance of an *emergency* are not necessarily *workers* prior to the *emergency*.

employer

A *person or organization* with recognized responsibilities, commitments and duties towards a *worker* in the employment of the *person or organization* by virtue of a mutually agreed relationship.

> ! A self-employed person is regarded as being both an *employer* and a *worker*.

end point

1.   The final stage of a *process*, especially the point at which an effect is observed.

> ⓘ Used to describe a range of different results or consequences. For example, the term 'biological *end point*' is used to describe a *health effect* (or a probability of that *health effect*) that could result from *exposure*.

2.   A radiological or other measure of *protection* or *safety* that is the calculated result of an *analysis* or *assessment*.

> ⓘ Common *end points* include estimates of *dose* or *risk*, estimated frequency or probability of an *event* or type of *event* (such as damage to the reactor core), expected number of *health effects* in a population, predicted environmental concentrations of radionuclides, etc.

3.   A predetermined criterion defining the point at which a specific task or *process* will be considered completed.

> ⓘ This usage often occurs in contexts such as *decontamination* or *remediation*, where the *end point* is typically the level of *contamination* beyond which further *decontamination* or *remediation* is considered unnecessary. One or more *end point* criteria are generally established for each *remedial action* or group of related *remedial actions* to verify their completion in accordance with the *remediation plan*.

> ⓘ In such a context, this criterion may also be an *end point* in the sense of definition (2) — such criteria are often calculated on the basis of a level of *dose* or *risk* that is considered acceptable — but its application to the actual *decontamination* or *remediation operations* is in the sense of definition (3).

end state

1.   The state of *radioactive waste* in the final stage of *radioactive waste management*, in which the *waste* is passively safe and does not depend on *institutional control*.

> ⓘ In the context of *radioactive waste management*, the *end state* refers to *disposal*.

2.   A predetermined criterion defining the point at which a specific task or *process* is to be considered completed.

> ⓘ Used in relation to *decommissioning activities* as the final state of *decommissioning* of a *facility*; and used in relation to *remediation* as the final status of a site at the end of *activities* for *decommissioning* and/or *remediation*, including approval of the radiological and physical conditions of the site and remaining *structures*.

> ⓘ Used in the context of probabilistic safety assessment, the end point is the set of conditions at the end of an event sequence that characterizes the impact of the sequence on the plant or the environment. Typically, two types of end state are specified: successful end state (e.g. no core damage) or unsuccessful end state (e.g. core damage, fuel damage, large release of radioactive material).

energy fluence

See *fluence*.

74

enforcement

The application by a *regulatory body* of sanctions against an *operator*, intended to correct and, as appropriate, penalize non-compliance with conditions of an *authorization*.

enriched uranium

See *uranium*.

entombment

See *decommissioning* (1).

entrance surface dose

Absorbed dose in the centre of the field at the surface of entry of *radiation* for a *patient* undergoing a radiodiagnostic *examination*, expressed in air and with backscatter.

environment

The conditions under which people, animals and plants live or develop and which sustain all life and development; especially such conditions as affected by human activities.

ⓘ See also *protection of the environment*.

environmental monitoring

See *monitoring* (1).

epicentre

The point on the Earth's surface directly above the focus (i.e. *hypocentre*) of an earthquake.

epistemic uncertainty

See *uncertainty*.

equilibrium, radioactive

See *radioactive equilibrium*.

equilibrium equivalent concentration (EEC)

The *activity concentration* of ^{222}Rn or ^{220}Rn in *radioactive equilibrium* with its short lived progeny that would have the same *potential alpha energy* concentration as the actual (non-equilibrium) mixture.

ⓘ The *equilibrium equivalent concentration* of ^{222}Rn is given by: $EEC\ ^{222}Rn = (0.104 \times C(^{218}Po)) + (0.514 \times C(^{214}Pb)) + (0.382 \times C(^{214}Bi))$ where $C(x)$ is the activity concentration of nuclide x in air. 1 Bq/m^3 $EEC\ ^{222}Rn$ corresponds to 5.56×10^{-6} mJ/m^3.

ⓘ The *equilibrium equivalent concentration* of ^{220}Rn is given by: $EEC\ ^{220}Rn = (0.913 \times C(^{212}Pb)) + (0.087 \times C(^{212}Bi))$ where $C(x)$ is the activity concentration of nuclide x in air. 1 Bq/m^3 $EEC\ ^{220}Rn$ corresponds to 7.57×10^{-5} mJ/m^3.

equilibrium factor

The ratio of the *equilibrium equivalent concentration* of ^{222}Rn to the actual ^{222}Rn activity concentration.

equipment qualification

See *qualification*.

equivalent dose

See *dose quantities*.

error management

Based on theories of perception, cognitive bias and anthropometry, this identifies the likelihood of errors made by humans in the system and technology interface.

  ⓘ *Human factors engineering* predicts errors and then designs to prevent the errors or their consequences from impacting the safe operation of the plant.

eruption, volcanic

See *volcanic eruption*.

eruption cloud

A cloud of tephra and gases that forms above a *volcanic vent* during explosive *volcanic eruptions*.

  ⓘ The vertical pillar of tephra and gases that forms during most explosive activity is referred to as an *eruption column*, or strong plume, and includes a momentum dominated region and a buoyancy dominated region.

  ⓘ *Eruption clouds* may rapidly spread laterally under gravity, especially in the most energetic eruptions, and may drift thousands of kilometres downwind.

  ⓘ Large *eruption clouds* can encircle the Earth within days.

essential services

  ⓘ The supply of resources, including electricity, gas, water, compressed air, fuel and lubricants, necessary to maintain *safety systems* of a nuclear power plant operational at all times.

evacuation

The rapid, temporary removal of people from an area to avoid or reduce short term *radiation exposure* in a *nuclear or radiological emergency*.

  ⓘ *Evacuation* is an *urgent protective action*. It is expected to be in place for a short period of time (e.g. a day to a few weeks). If *evacuation* cannot be lifted within this short period of time, it should be substituted by *relocation*.

  ⓘ *Evacuation* may be taken as a *precautionary urgent protective action* based on observable conditions or plant conditions.

  See also *relocation*.

event

In the context of the reporting and *analysis* of *events*, an *event* is any occurrence unintended by the *operator*, including operating error, equipment *failure* or other mishap, and deliberate action on the part of others, the consequences or potential consequences of which are not negligible from the point of view of *protection and safety*.

! The terminology related to the reporting and *analysis* of *events* is not consistent with the terminology used in *safety standards*, and great care should be taken to avoid confusion.

! In particular, the definition of *event* given above is identical in essence to the *safety standards* definition (1) of *accident*.

ⓘ This difference derives from the fact that *event* reporting and *analysis* is concerned directly with the question of whether an *event* that could develop into an *accident* with significant consequences actually does so; terms such as *accident* are used only to describe the end result and therefore other terms are needed to describe the earlier stages.

See also *initiating event* and *initiating event: postulated initiating event*.

ⓘ *Event* is also used in the phrase 'features, *events* and *processes*' associated with the site and the *facility* in the context of *site characterization* for a *disposal facility* for *radioactive waste*.

ⓘ Relevant features, *events* and *processes* relating to the site are those that might influence the long term performance of the *disposal facility* and thus could affect *safety*. These are addressed in a *safety case* and in a supporting *safety assessment*.

TABLE 1. TYPES OF EVENTS AND CIRCUMSTANCES

Events (including *anticipated operational occurrences*)			Circumstances		
Incidents (including *initiating events*, *accident precursors* and *near misses*)		*Scenarios:* postulated incidents	Situations (including *operating conditions, accident conditions*)		*Scenarios:* hypothetical situations
Accidents (unintentional causes)	Intentional causes (unauthorized acts: malicious and non-malicious) (e.g. sabotage, theft)	E.g. acute *potential exposure*	*Operational states, design basis accident conditions*	Nuclear and radiological emergencies, beyond design basis accident conditions	E.g. chronic *potential exposure*

Notes: A *scenario* is a postulated or assumed set of conditions and/or *events*. A *scenario* may represent the conditions at a single point in time or a single *event*, or a time history of conditions and/or *events*.

Anticipated operational occurrence; *beyond design basis accident*; *design basis accident*: see *plant states (considered in design)*.

These terms use the following attributes: acute and chronic; actual and postulated; unintentional and intentional causes; malicious and non-malicious; *nuclear* and radiological.

Dictionary (Concise Oxford English Dictionary [34]) definitions:

Circumstance: A fact or condition connected with or relevant to an *event* or an action.

Occurrence: The fact or frequency of something occurring; an *incident* or *event*.

Situation: A set of circumstances in which one finds oneself.

event tree analysis

See *analysis*.

examination

A procedure used to obtain information from evidence in order to reach conclusions concerning the nature of and/or associations related to evidence.

    ⓘ This definition is for use in the context of *nuclear security*.

    ⓘ Should not normally need a definition: if used, the text should make clear the specific meaning in the context of the publication.

exception

    ⓘ The terms *exception* and excepted are sometimes used to describe cases in which some of the *requirements* or guidance in *safety standards* are deemed not to apply.

    ⓘ In this regard, the effect of *exception* may be compared with the effects of *exemption* and *exclusion*.

    ⓘ However, this is in fact a usual usage of the English term *exception*, not a specialized term.

    ⓘ The terms *exemption* and *exclusion* are necessarily linked to specific reasons for non-application, whereas *exception* is not.

    ⓘ The term 'excepted *package*' in the Transport Regulations [2] is an example of this usage; *packages* may be excepted from specified *requirements* of the Transport Regulations if they satisfy conditions specified in the Transport Regulations.

excess risk

See *risk* (3).

excluded exposure

Exposure not considered amenable to control through a regulatory instrument.

    ⓘ The term *excluded exposure* is most commonly applied to those *exposures* due to *natural sources* that are least amenable to *control*, such as *exposures* due to cosmic *radiation* at the Earth's surface, ^{40}K in the human body or *naturally occurring radioactive material (NORM)* in which the *activity concentrations* of natural radionuclides are below the relevant values given in IAEA *safety standards*.

ⓘ The concept is related to those of *clearance* (which is normally used in relation to materials) and *exemption* (which relates to *facilities and activities* or *sources*).

ⓘ See also *exclusion*.

exclusion

The deliberate excluding of a particular type of *exposure* from the scope of an instrument of *regulatory control* on the grounds that it is not considered amenable to *control* through the regulatory instrument in question.

exclusive use

The sole use, by a single *consignor*, of a *conveyance* or of a *large freight container*, in respect of which all initial, intermediate and final loading and unloading and *shipment* are carried out in accordance with the directions of the *consignor* or *consignee*, where so required by [the Transport] Regulations. (See SSR-6 (Rev. 1) [2].)

exempt waste

See *waste*.

exemption

The determination by a *regulatory body* that a *source* or *practice* need not be subject to some or all aspects of *regulatory control* on the basis that the *exposure* and the *potential exposure* due to the *source* or *practice* are too small to warrant the application of those aspects or that this is the optimum option for *protection* irrespective of the actual level of the *doses* or *risks*.

ⓘ For *exemption* from IAEA safeguards, see the Safeguards Glossary [14].

See also *clearance* (1) and *exclusion*.

exemption level

See *level*.

existing exposure situation

See *exposure situations*.

explosive eruption

See volcanic *eruption*.

exposure

1.   The state or condition of being subject to irradiation.

> ! *Exposure* should not be used as a synonym for *dose*. *Dose* is a measure of the effects of *exposure*.

> ⓘ *Exposure* to *ionizing radiation* can be broadly divided into *exposure categories* according to the status of the individual(s) exposed; into *exposure situations* according to the circumstances of the exposure; and according to the *source* of the *exposure*.

acute exposure. *Exposure* received within a short period of time.

> ⓘ Normally used to refer to *exposure* of sufficiently short duration that the resulting *doses* can be treated as instantaneous (e.g. less than an hour).

external exposure. *Exposure* to *radiation* from a *source* outside the body.

> ⓘ Contrasted with *internal exposure*.

internal exposure. *Exposure* to *radiation* from a *source* within the body.

> ⓘ Contrasted with *external exposure*.

transboundary exposure. *Exposure* of *members of the public* in one State due to *radioactive material* released via *accidents*, *discharges* or *waste disposal* in another State.

> ⓘ See also *potential exposure*.

2.   The sum of the electrical charges of all the ions of one sign produced in air by X rays or gamma *radiation* when all electrons liberated by photons in a suitably small element of volume of air are completely stopped in air, divided by the mass of the air in the volume element.

> ⓘ Unit: C/kg (in the past, the *röntgen (R)* was used).

3.   The time integral of the *potential alpha energy* concentration in air, or of the corresponding *equilibrium equivalent concentration*, to which an individual is exposed over a given period (e.g. a year).

> ⓘ Used in connection with *exposure* due to decay products of ^{222}Rn or ^{220}Rn.

> ⓘ The SI unit is J·h/m^3 for *potential alpha energy* concentration or Bq·h/m^3 for *equilibrium equivalent concentration*.

exposure due to radon. The time integral over the *activity concentration* of *radon* for a defined period of time. *Exposure due to radon* is a measurand related to the *potential alpha energy exposure* with the *equilibrium factor* taken into account and is, therefore, related to the *effective dose*.

4.   ["[T]he product of the air concentration of a radionuclide to which a person is exposed … and the time of exposure. More generally, when the air concentration varies with time, the time integral of the air concentration of a radionuclide to which a person is exposed, integrated over the time of exposure."]

> ⓘ This definition, quoted verbatim from Ref. [35], reflects a loose usage of exposure found in particular in the context of airborne *radon*. This usage is listed here for information, but it is discouraged.

exposure assessment

See *assessment* (1): *dose assessment*.

exposure categories

medical exposure. *Exposure* incurred by *patients* for the purposes of their own medical or dental diagnosis (**diagnostic exposure**) or medical treatment (**therapeutic exposure**); by *carers and comforters*; and by volunteers subject to *exposure* as part of a programme of biomedical research.

ⓘ See *patient.*

occupational exposure. *Exposure* of *workers* incurred in the course of their work.

public exposure. *Exposure* incurred by *members of the public* due to *sources* in *planned exposure situations, emergency exposure situations* and *existing exposure situations*, excluding any *occupational exposure* or *medical exposure*.

exposure due to radon

See *exposure* (3).

exposure pathway

A route by which *radiation* or radionuclides can reach humans and cause *exposure*.

ⓘ An *exposure pathway* may be very simple, for example the external *exposure pathway* from airborne radionuclides, or a more complex chain, for example the *internal exposure pathway* from drinking milk from cows that ate grass contaminated with deposited radionuclides.

exposure situations

! The *exposure situation* is indicated by the circumstances of *exposure* of the individual(s) undergoing *exposure*; it cannot be used to characterize a jurisdiction or the geographical area, for example, although for practical purposes such generalizations are sometimes assumed.

ⓘ Three broad *exposure situations* were used as the basis for organizing the *safety requirements* established in GSR Part 3 [1]. The characterizations in terms of 'situations' (which derive from Ref. [33]) are not clearly delineated or conceptually distinct and the descriptions of the three types of *exposure situation* are not always sufficient to determine unequivocally which type of *exposure situation* applies for particular circumstances. In the *safety standards*, the most appropriate type of *exposure situation* for particular circumstances is determined by taking practical considerations into account.

emergency exposure situation. A situation of *exposure* that arises as a result of an *accident*, a *malicious act* or other unexpected event, and requires prompt action in order to avoid or to reduce adverse consequences.

ⓘ *Exposure* in an *emergency* can include both *occupational exposure* and *public exposure*, and can include unplanned *exposures* resulting directly in the *emergency exposure situation* and planned *exposures* to *emergency workers* and *helpers in an emergency* undertaking actions to mitigate the consequences of the *emergency*.

ⓘ *Exposure* in an *emergency* can be reduced only by *protective actions* and *other response actions*.

existing exposure situation. A situation of *exposure* that already exists when a decision on the need for *control* needs to be taken.

ⓘ *Existing exposure situations* include exposure to *natural background* radiation that is amenable to *control*; exposure due to residual *radioactive material* that derives from past *practices* that were never subject to *regulatory control*; and exposure due to residual *radioactive material* deriving from a *nuclear or radiological emergency* after an *emergency* has been declared to be ended. See para. 5.1 and Requirement 52 of GSR Part 3 [1].

planned exposure situation. A situation of *exposure* that arises from the planned operation of a *source* or from a planned *activity* that results in an *exposure* due to a *source*.

   ⓘ Since provision for *protection and safety* can be made before embarking on the *activity* concerned, associated *exposures* and their probabilities of occurrence can be restricted from the outset.

   ⓘ The primary means of controlling *exposure* in *planned exposure situations* is by good *design* of installations, equipment and operating procedures. In *planned exposure situations*, a certain level of exposure is expected to occur.

extended planning distance (EPD)

See *emergency planning distance*.

external adversary

See *adversary*.

external event

Events unconnected with the *operation* of a *facility* or the conduct of an *activity* that could have an effect on the *safety* of the *facility* or *activity*.

   ⓘ Typical examples of *external events* for *nuclear facilities* include earthquakes, tornadoes, tsunamis and aircraft crashes.

   ⓘ In the case of *safety assessment* for long term *safety* in *waste management*, a relevant *external event* is one that could have an effect on the functioning of *multiple barriers*.

external exposure

See *exposure* (1).

external hazard

See hazard.

external zone

The area immediately surrounding a proposed *site area* in which population distribution and density, and land and water uses, are considered with respect to their impact on planning effective *emergency response actions*.

   ⓘ Used in the context of *siting* of *facilities*.

   ⓘ This is the area that would be the *emergency planning zones* if the *facility* were in place.

F

facilities and activities

A general term encompassing *nuclear facilities*, uses of all *sources* of *ionizing radiation*, all *radioactive waste management activities*, *transport* of *radioactive material* and any other *practice* or circumstances in which people may be subject to *exposure* to *radiation* from naturally occurring or artificial *sources*.

ⓘ *'Facilities'* includes: *nuclear facilities*; *irradiation installations*; some mining and raw material processing *facilities* such as *uranium* mines; *radioactive waste management facilities*; and any other places where *radioactive material* is produced, processed, used, handled, stored or disposed of — or where *radiation generators* are installed — on such a scale that consideration of *protection and safety* is required.

ⓘ *'Activities'* includes: the production, use, import and export of *radiation sources* for industrial, research and medical purposes; the *transport* of *radioactive material*; the *decommissioning* of *facilities*; *radioactive waste management activities* such as the *discharge* of effluents; and some aspects of the *remediation* of sites affected by residues from past *activities*.

ⓘ The intention is to include any human *activity* that introduces additional *sources* of *radiation* or additional *exposure pathways*, or that modifies the network of *exposure pathways* from existing *sources*, so as to increase the *exposure* or the likelihood of *exposure* of people or the number of people exposed.

ⓘ The term *'facilities and activities'* is intended to provide an alternative to the terminology of *sources* and *practices* (or *interventions*) to refer to general categories of situations.

ⓘ For example, a *practice* may involve many different *facilities and/or activities*, whereas the general definition (1) of *source* is too broad in some cases: a *facility or activity* might constitute a *source*, or might involve the use of many *sources*, depending upon the interpretation used.

ⓘ The term *'facilities and activities'* is very general, and includes those for which little or no *regulatory control* may be necessary or achievable: the more specific terms **authorized facility** and **authorized activity** should be used to distinguish those *facilities and activities* for which any form of *authorization* has been given.

ⓘ In the Fundamental Safety Principles (Safety Fundamentals), the term *'facilities and activities* — existing and new — utilized for peaceful purposes' is abbreviated for convenience to *facilities and activities* as a general term encompassing any human activity that may cause people to be exposed to *radiation risks* arising from naturally occurring or artificial *sources* (see SF-1 [24], para. 1.9).

ⓘ For safeguards purposes, see the definition of *facility* in the Safeguards Glossary [14].

ⓘ *Facilities and activities* are listed as follows in GSR Part 4 (Rev. 1) [19]:

'*Facilities*' includes:

(a)   Nuclear power plants;

(b)   Other reactors (such as *research reactors* and *critical assemblies*);

(c)   Enrichment *facilities* and *nuclear fuel* fabrication facilities;

(d)   Conversion facilities used to generate *uranium* hexafluoride (UF_6);

(e)   *Storage* facilities and *reprocessing* plants for irradiated fuel;

(f)   *Facilities* for *radioactive waste management* where *radioactive waste* is treated, conditioned, stored or disposed of;

(g)   Any other places where *radioactive materials* are produced, processed, used, handled or stored;

(h)   *Irradiation installations* for medical, industrial, research and other purposes, and any places where *radiation generators* are installed;

(i)   *Facilities* where the mining and processing of *radioactive* ores (such as ores of *uranium* and thorium) are carried out.

'*Activities*' includes:

(a)   The production, use, import and export of *radiation sources* for medical, industrial, research and other purposes;

(b)   The *transport* of *radioactive material*;

(c)   The *decommissioning* of *facilities* and the *closure* of *repositories* for *radioactive waste*;

(d)   The close-out of *facilities* where the mining and processing of *radioactive* ores was carried out;

(e)   *Activities* for *radioactive waste management* such as the *discharge* of effluents;

(f)   The *remediation* of sites affected by residues from past *activities*.

associated activity. The possession, production, processing, use, handling, storage, disposal or transport of *nuclear material* or *other radioactive material*.

!   Although the wording does not explicitly exclude malicious activities conducted by *adversaries*, this term is presumably intended to refer only to *authorized activities*.

§   This term is broadly equivalent to an '*activity*' in the general term *facilities and activities*.

associated facility. A facility (including associated buildings and equipment) in which *nuclear material* or *other radioactive material* is produced, processed, used, handled, stored or disposed of and for which an *authorization* is required.

ⓘ   This includes *nuclear facilities* and any other facilities holding significant amounts of *radioactive material*.

§   This term is broadly equivalent to a '*facility*' in the general term *facilities and activities*.

facility

See *facilities and activities*.

facility states (considered in design)

ⓘ The concept of *facility states* as it is used in the *safety standards* for *research reactors* and for *nuclear fuel cycle facilities* is broadly equivalent to the concept of *plant states* for nuclear power plants. See *plant states (considered in design)* for related terms and definitions (namely *operational states, normal operation, anticipated operational occurrences, accident conditions, design basis accident, design extension conditions, controlled state, safe state*); see also *plant equipment (for a nuclear power plant): safety features for design extension conditions*.

1.   **facility states** (postulated states of a *research reactor* facility as considered for design purposes).

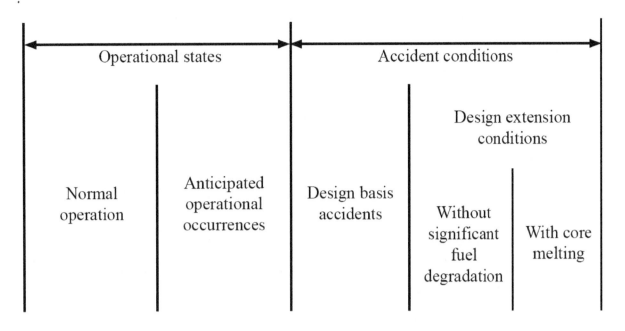

FIG. 1. Facility states considered in design for a research reactor.

See SSR-3 [36].

2.   **facility states** (postulated states of a *nuclear fuel cycle facility* as considered for design purposes).

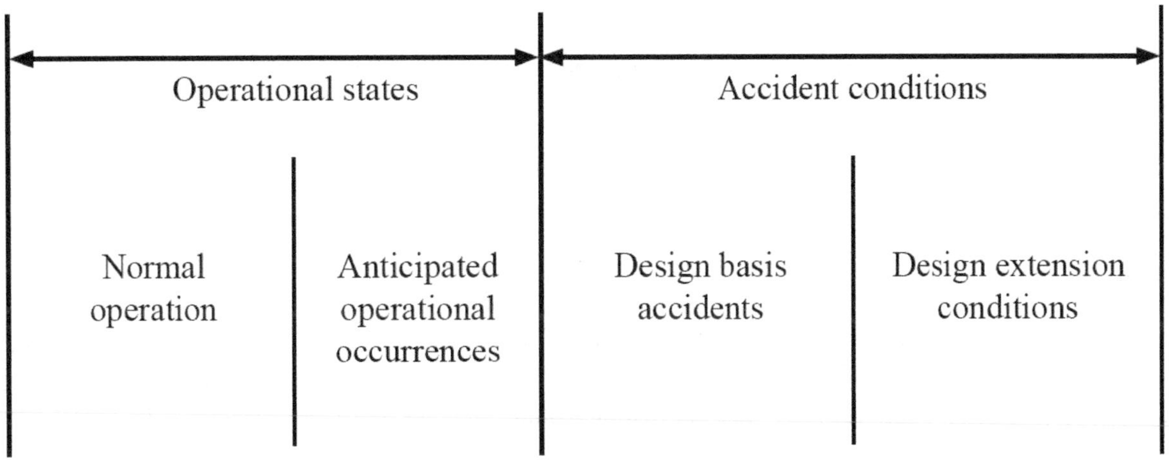

FIG. 2. Facility states considered in design for a nuclear fuel cycle facility.

See SSR-4 [37].

facility emergency

See *emergency class.*

failure (technical)

Loss of the ability of a *structure, system or component* to function within *acceptance criteria*.

> ! Note that the *structure, system or component* is considered to fail when it becomes incapable of functioning, whether or not this is needed at that time.

> ! A *failure* in, for example, a backup *system* may not be manifest until the *system* is called upon to function, either during testing or on *failure* of the *system* it is backing up.

> ⓘ A *failure* may be the result of, for example, a hardware fault, a software fault, a system fault, an operator error or a maintenance error.

> **common cause failure.** *Failures* of two or more *structures, systems or components* due to a single specific *event* or cause.

> ⓘ For example, the single specific *event* or cause of *failures* (which may be *failures* of different types) could be a *design* deficiency, a manufacturing deficiency, *operation* and *maintenance* errors, a natural phenomenon, a human induced *event*, saturation of signals, or an unintended cascading effect from any other *operation* or *failure* within the plant or from a change in ambient conditions.

> ⓘ Common causes may be internal or external to a system.

> **common mode failure.** *Failures* of two or more *structures, systems or components* in the same manner or mode due to a single specific *event* or cause.

> ⓘ *Common mode failure* is a type of *common cause failure* in which the *structures, systems or components* fail in the same way (although they may not be in close proximity).

failure mode

The manner or state in which a *structure, system or component* fails.

false alarm

See *alarm*.

far field

The *geosphere* outside a *disposal facility*, comprising the surrounding geological strata, at a distance from the *disposal facility* such that, for modelling purposes, the *disposal facility* may be considered a single entity and the effects of individual *waste packages* are not distinguished.

> ⓘ For practical purposes, this is often interpreted simply as the *geosphere* beyond the *near field*.

fault, geological

See *geological fault*.

fault tree analysis

See *analysis*.

feed

Any single material or multiple materials, whether processed, semi-processed or raw, that is or are intended to be fed directly to *food* producing animals.

field programmable gate array

An integrated circuit that can be programmed in the field by the instrumentation and control manufacturer.

> ⓘ A *field programmable gate array* includes programmable *logic* blocks (combinatorial and sequential), programmable interconnections between them and programmable blocks for input and/or outputs. The function is then defined by the instrumentation and control designer, not by the circuit manufacturer.

fire barrier

See *barrier*.

fire load

The sum of the calorific energies calculated to be released by the complete combustion of all the *combustible materials* in a space, including the facings of the walls, partitions, floors and ceiling.

firmware

Software which is closely coupled to the hardware characteristics on which it is installed.

first responders

The first members of an *emergency service* to respond at the site of an *emergency*.

fissile material

1.   Material containing any of the *fissile nuclides (1)* in sufficient proportion to enable a self-sustained nuclear chain reaction with slow (thermal) neutrons.

2.   Material containing any of the *fissile nuclides*. Excluded from the definition of *fissile material* are the following:

   (a)   *Natural uranium* or *depleted uranium* that is *unirradiated*;

   (b)   *Natural uranium* or *depleted uranium* that has been irradiated in thermal reactors only;

   (c)   Material with *fissile nuclides* less than a total of 0.25 g;

   (d)   Any combination of (a), (b) and/or (c).

These exclusions are valid only if there is no other material with *fissile nuclides* in the *package* or in the *consignment* if shipped unpackaged. (See SSR-6 (Rev. 1) [2].)

> ⓘ Definition (2) is specific to the Transport Regulations [2]. As with *radioactive material*, this is not a scientific definition, but one designed to serve a specific regulatory purpose.

See also *fissionable material*.

fissile nuclide

1.   Nuclides, in particular ^{233}U, ^{235}U, ^{239}Pu and ^{241}Pu, that are able to support a self-sustaining nuclear chain reaction with neutrons of all energies, but predominantly with slow neutrons.

2.   Uranium-233, ^{235}U, ^{239}Pu and ^{241}Pu.

> ⓘ Definition (2) is specific to the Transport Regulations [2]. As with *radioactive material*, this is not a scientific definition, but one designed to serve a specific regulatory purpose.

fission fragment

A nucleus resulting from nuclear fission carrying kinetic energy from that fission.

> ⓘ Used only in contexts where the particles themselves have kinetic energy and thus could represent a *hazard*, irrespective of whether the particles are *radioactive*.

> ⓘ Otherwise, the more usual term *fission product* is used.

fission product

A radionuclide produced by nuclear fission.

> ⓘ Used in contexts where the *radiation* emitted by the radionuclide is the potential *hazard*.

fissionable material

Material containing any *fissionable nuclides*.

> *fissionable nuclide.* Nuclides that are capable of supporting a self-sustaining nuclear chain reaction with neutrons of any speed.

See also *fissile material*.

fissionable nuclide

See *fissionable material*.

fixed contamination

See *contamination* (2).

fluence

> ⓘ A measure of the strength of a *radiation* field. Commonly used without qualification to mean *particle fluence*.

> *energy fluence, Ψ.* A measure of the energy density of a *radiation* field, defined as:

$$\Psi = \frac{dR}{da}$$

where dR is the *radiation* energy incident on a sphere of cross-sectional area da.

① The *energy fluence* rate

$$\frac{\mathrm{d}\Psi}{\mathrm{d}t}$$

is denoted by a lower case ψ.

See Ref. [38].

particle fluence, **Φ.** A measure of the density of particles in a *radiation* field, defined as:

$$\Phi = \frac{\mathrm{d}N}{\mathrm{d}a}$$

where $\mathrm{d}N$ is the number of particles incident on a sphere of cross-sectional area $\mathrm{d}a$.

① The *particle fluence* rate

$$\frac{\mathrm{d}\Phi}{\mathrm{d}t}$$

is denoted by a lower case φ.

See Ref. [38].

food

Any substance, whether processed, semi-processed or raw, that is intended for human consumption.

① This includes foodstuffs and drink (other than fresh water), chewing gum and substances used in the preparation or processing of *food*; it does not include cosmetics, tobacco or drugs. Consumption in this context refers to ingestion.

force-on-force exercise

A *performance test* of the *physical protection system* that uses designated trained personnel in the role of an *adversary* force to simulate an attack consistent with the *threat* or the *design basis threat*.

fractional absorption in the gastrointestinal tract, f_1, or in the alimentary tract, f_A

The fraction of an ingested element that is directly absorbed to body fluids. (See Refs [27–29, 39].)

① Often referred to colloquially as *gut transfer factor* or 'f_1 value'.

See also *lung absorption type*, a similar concept for *activity* in the respiratory tract.

free field ground motion

Motion that would occur at a given point on the ground owing to an earthquake if vibratory characteristics were not affected by *structures* and *facilities*.

freight container

An article of *transport* equipment that is of a permanent character and accordingly strong enough to be suitable for repeated use; specially designed to facilitate the *transport* of goods by one or other modes of *transport* without intermediate reloading, designed to be secured and/or readily handled, having fittings for these purposes.

ⓘ The *freight container* does not include the *vehicle*.

small freight container. A *freight container* that has an internal volume of not more than 3 m³. (See SSR-6 (Rev. 1) [2].)

large freight container. A *freight container* that has an internal volume of more than 3 m³. (See SSR-6 (Rev. 1) [2].)

frequency of exceedance

The frequency at which a specified level of seismic *hazard* will be exceeded at a site or in a region within a specified time interval.

ⓘ In probabilistic seismic *hazard* analysis, generally a one year time interval (i.e. annual frequency) is assumed.

ⓘ When the frequency is very small and it cannot exceed unity (in the prescribed interval), this number approaches the probability of the same event if the random process is assumed to be Poissonian.

fresh fuel

See *nuclear fuel*.

fuel

See *nuclear fuel*.

fuel assembly

A set of *fuel elements* and associated *components* which are loaded into and subsequently removed from a reactor core as a single unit.

fuel cycle

See *nuclear fuel cycle*.

fuel element

A rod of *nuclear fuel*, its *cladding* and any associated *components* necessary to form a structural entity.

ⓘ Commonly referred to as a *fuel rod* in light water reactors.

fuel rod

See *fuel element*.

functional diversity

See *diversity*.

functional indicator

See *indicator*.

functional isolation

Prevention of adverse consequences from the mode of *operation* or *failure* of one circuit or *system* on another.

functional requirements

Requirements that specify the required functions or behaviours of an item.

> ***non-functional requirements.*** Requirements that specify inherent properties or characteristics of an item other than the required functions and behaviours.
>
> ⓘ Example characteristics include analysability, assurability, auditability, *availability*, compatibility, documentation, integrity, maintainability, reliability, *safety*, *security*, usability and verifiability.
>
> ⓘ Also known as quality requirements.

fundamental safety function

See *safety function*.

G

gap release

Release, especially in a reactor core, of *fission products* from the *fuel* pin gap, which occurs immediately after *failure* of the *fuel cladding* and is the first radiological indication of *fuel* damage or *fuel failure*.

general emergency

See *emergency class*.

generic criteria

Levels for the *projected dose* or the *dose* that has been received at which *protective actions* and *other response actions* are to be taken.

ⓘ The term *generic criteria* as defined here relates to *emergency preparedness* and *emergency response* only.

genetic effect

See *health effects (of radiation)*: *hereditary effect*.

geological disposal

See *disposal* (1).

geological disposal facility

See *disposal facility*.

geological fault

A planar or gently curved fracture surface or zone of the Earth across which there has been relative displacement.

capable fault. A *geological fault* that has a significant potential for displacement at or near the ground surface.

ⓘ A *geological fault* is to be considered a *capable fault* if, on the basis of geological, geophysical, geodetic or seismological data (including paleoseismological and geomorphological data), one or more of the following conditions applies:

(a) The *geological fault* shows evidence of past movement or movements (significant deformations and/or dislocations) of a recurring nature within such a period that it is reasonable to infer that further movements at or near the surface could occur.

(b) A structural relationship with a known *capable fault* has been demonstrated such that movement of the one may cause movement of the other at or near the surface.

(c) The maximum potential earthquake associated with a seismogenic structure is sufficiently large and at such a depth that it is reasonable to infer that, in the geodynamic setting of the site, movement at or near the surface could occur [40].

ⓘ In highly active areas, where both earthquake data and geological data consistently reveal short earthquake recurrence intervals, periods of the order of tens of thousands of years may be appropriate for the assessment of *capable faults*. In less active areas, it is likely that much longer periods may be required.

geological record

The sequence of rock layers in a vertical section of the Earth.

ⓘ Also termed the stratigraphic record. The oldest layers occur at the base of the section, with successively younger layers occurring higher in the sequence.

ⓘ Geologists use the geological record to assign relative ages to deposits.

ⓘ Volcanic stratigraphy is often complex, with deposits characterized by having relatively limited lateral extent, exhibiting rapid facies changes and having undergone multiple episodes of erosion and refilling of valleys.

geosphere

Those parts of the lithosphere not considered to be part of the *biosphere*.

ⓘ In *safety assessment*, usually used to distinguish the subsoil and rock (below the depth affected by usual human *activities*, in particular agriculture) from the soil that is part of the *biosphere*.

grace period

The period of time during which a *safety function* is ensured in an *event* with no necessity for action by personnel.

ⓘ Typical *grace periods* range from 20 min to 12 h. The *grace period* may be achieved by means of the automation of actuations, the adoption of passive *systems* or the inherent characteristics of a material (such as the heat capacity of the *containment structure*), or by any combination of these.

graded approach

1.   For a system of *control*, such as a regulatory system or a *safety system*, a *process* or method in which the stringency of the *control* measures and conditions to be applied is commensurate, to the extent practicable, with the likelihood and possible consequences of, and the level of *risk* associated with, a loss of *control*.

ⓘ An example of *a graded approach* in general would be a structured method by means of which the stringency of application of *requirements* is varied in accordance with the circumstances, the regulatory systems used, the *management systems* used, etc.

ⓘ For example, a method in which:

(1) The significance and complexity of a product or service are determined;

(2) The potential impacts of the product or service on health, *safety*, *security*, the *environment*, and the achieving of quality and the organization's objectives are determined;

(3) The consequences if a product fails or if a service is carried out incorrectly are taken into account.

ⓘ The use of a *graded approach* is intended to ensure that the necessary levels of *analysis*, documentation and actions are commensurate with, for example, the magnitudes of any radiological *hazards* and non-radiological *hazards*, the nature and the particular characteristics of a *facility*, and the stage in the *lifetime* of a *facility*.

2.   An application of *safety requirements* that is commensurate with the characteristics of the *facilities and activities* or the *source* and with the magnitude and likelihood of the *exposures*.

See also *exclusion, exemption, clearance* and *optimization*.

3.   The application of *nuclear security measures* proportionate to the potential consequences of a *malicious act*.

> ⓘ In some Nuclear Security Series publications, this term has also been defined in a broadly similar way as "The application of *nuclear security measures* proportionate [or proportional] to the potential consequences of criminal or intentional unauthorized acts involving or directed at *nuclear material, other radioactive material*, associated facilities or associated activities or other acts determined by the State to have an adverse impact on nuclear security" [8, 9].

gray (Gy)

The SI unit of *kerma* and *absorbed dose*, equal to 1 J/kg.

ground shine

Gamma *radiation* from radionuclides deposited on the ground.

> ⓘ *Ground shine* is of concern as an *exposure pathway* for *external exposure* principally — but not exclusively — to gamma *radiation*.

> ⓘ *Ground shine* may also be used to mean *radiation* that is incident on, and reflected back from, the ground.

See also *cloud shine*.

guard

A person who is entrusted with responsibility for patrolling, monitoring, assessing, escorting individuals or *transport*, controlling access and/or providing initial response.

> ⓘ People carrying out one of these tasks, e.g. escorting an individual, are not necessarily *guards*.

gut transfer factor

See *fractional absorption in the gastrointestinal tract, f_1, or in the alimentary tract, f_A*.

H

habit survey

See *survey*.

half-life, $T_{1/2}$

1.   For a radionuclide, the time required for the *activity* to decrease, by a *radioactive* decay *process*, by half.

    ⓘ Where it is necessary to distinguish this from other *half-lives* (see (2)), the term *radioactive half-life* should be used.

    ⓘ The *half-life* is related to the *decay constant, λ*, by the expression:

$$T_{1/2} = \frac{\ln 2}{\lambda}$$

2.   The time taken for the quantity of a specified material (e.g. a radionuclide) in a specified place to decrease by half as a result of any specified *process* or *processes* that follow exponential patterns similar to *radioactive* decay.

    biological half-life. The time taken for the quantity of a material in a specified tissue, organ or region of the body (or any other specified biota) to halve as a result of biological *processes*.

    effective half-life, T_{eff}. The time taken for the *activity* of a radionuclide in a specified place to halve as a result of all relevant *processes*.

$$\frac{1}{T_{eff}} = \sum_i \frac{1}{T_i}$$

    where T_i is the *half-life* for *process* i.

    radioactive half-life. For a radionuclide, the time required for the *activity* to decrease, by a *radioactive* decay *process*, by half.

    ⓘ The term ***physical half-life*** is also used for this concept.

(harmful) tissue reaction

See *health effects (of radiation)*: *deterministic effect*: *severe deterministic effect*.

hardware description language

Language that allows one to formally describe the functions and/or the structure of an electronic *component*, for documentation, simulation or synthesis.

hardware programmed device

An integrated circuit configured (for instrumentation and control systems in nuclear power plants), with hardware description languages and related software tools.

harsh environment

Environmental conditions that are significantly more severe than the conditions anticipated for *operational states*.

hazard

The potential for harm or other detriment, especially for *radiation risks*; a factor or condition that might operate against *safety*.

> ***contributory hazard.*** Factor contributing to potential for harm.

> ***external hazard.*** *Hazard* that originates from outside the *site boundary* and outside the *activities* that are under the *control* of the *operating organization*, over which the *operating organization* has very little or no *control*, but that could have an effect on the *safety* of the *facility or activity*.

> ***internal hazard.*** *Hazard* to the *safety* of a *facility* that originates from within the *site boundary* and is associated with failures of *facilities and activities* that are under the *control* of the *operating organization*.

hazard assessment

See *assessment* (1).

hazard analysis

See *analysis*.

health authority

A governmental authority (at the national, regional or local level) that is responsible for policies and interventions, including the development of standards and the provision of guidance, for maintaining or improving human health, and that has the legal power of enforcing such policies and interventions.

health effects (of radiation)

> ***deterministic effect.*** A *radiation* induced *health effect* for which generally a threshold level of *dose* exists above which the severity of the effect is greater for a higher *dose*.

> > ***severe deterministic effect.*** A *deterministic effect* that is fatal or life threatening or results in a permanent injury that reduces quality of life.

> ⓘ The level of the threshold *dose* is characteristic of the particular *health effect* but may also depend, to a limited extent, on the exposed individual.

> ⓘ Examples of *deterministic effects* include erythema, damage to the haemopoietic system and acute *radiation* syndrome (*radiation* sickness).

> ⓘ *Deterministic effects* are also referred to as ***(harmful) tissue reactions***.

> ⓘ The term **[*non-stochastic effect*]** is used in some older publications, but is now superseded.

> ⓘ Contrasting term: *stochastic effect*.

early effect. A *radiation* induced *health effect* that occurs within months of the *exposure* that caused it.

ⓘ All *early effects* are *deterministic effects*; most, but not all, *deterministic effects* are *early effects*.

hereditary effect. A *radiation* induced *health effect* that occurs in a descendant of the exposed person.

ⓘ The less precise term **genetic effect** is also used, but *hereditary effect* is preferred.

ⓘ *Hereditary effects* are usually *stochastic effects*.

ⓘ Contrasting term: *somatic effect*.

late effect. A *radiation* induced *health effect* that occurs years after the *exposure* that caused it.

ⓘ The most common *late effects* are *stochastic effects*, such as leukaemia and solid cancers, but some *deterministic effects* can also be *late effects*.

somatic effect. A *radiation* induced *health effect* that occurs in the exposed person.

ⓘ This includes effects occurring after birth that are attributable to *exposure* in utero.

ⓘ *Deterministic effects* are normally also *somatic effects*; *stochastic effects* may be *somatic effects* or *hereditary effects*.

ⓘ Contrasting term: *hereditary effect*.

stochastic effect. A *radiation* induced *health effect*, the probability of occurrence of which is greater for a higher *radiation dose* and the severity of which (if it occurs) is independent of *dose*.

ⓘ *Stochastic effects* may be *somatic effects* or *hereditary effects* and generally occur without a threshold level of *dose*. Examples include solid cancers and leukaemia.

ⓘ Contrasting term: *deterministic effect*.

health professional

An individual who has been formally recognized through appropriate national *procedures* to practise a profession related to health (e.g. medicine, dentistry, chiropractic, podiatry, nursing, medical physics, medical radiation technology, radiopharmacy, occupational health).

ⓘ Used to distinguish from a *referring medical practitioner* or a *radiological medical practitioner*, who satisfy additional criteria.

health screening programme

A programme in which health tests or medical *examinations* are performed for the purpose of early detection of disease.

health surveillance

See *workers' health surveillance*.

[heat generating waste (HGW)]

See *waste classes*.

helper in an emergency

Member of the public who willingly and voluntarily helps in the response to a *nuclear or radiological emergency.*

> ⓘ *Helpers in an emergency* are protected and are aware that they could be exposed to *radiation* while helping in response to a *nuclear or radiological emergency.*

hereditary effect

See *health effects (of radiation)*.

high confidence of low probability of failure

The earthquake level for which there is 95% confidence that the probability of failure of a structure, system or component is less than 5%.

> ⓘ It also represents the acceleration corresponding to the mean fragility of the 1% conditional probability of failure. *High confidence of low probability of failure* is a measure of the seismic margin capacity of a *structure, system or component.*

high energy radiation therapy equipment

X ray equipment and other types of *radiation generators* capable of operating at generating potentials above 300 kV, and radionuclide teletherapy equipment.

high enriched uranium (HEU)

See *uranium*.

high level waste (HLW)

See *waste classes*.

high linear energy transfer (LET) radiation

See *radiation*.

Holocene

The most recent epoch of the geological Quaternary period, defined as the interval from 10 000 years before the present to the present.

Holocene volcano

See *volcano*.

human factors engineering

Engineering in which factors that could influence human performance and that could affect *safety* are understood and are taken into account, especially in the *design* and *operation* of *facilities*.

> See also *error management, important human task*.

human intrusion

ⓘ The term *human intrusion* is used for human *activities* that could affect the integrity of a *disposal facility* and which could potentially give rise to radiological consequences.

ⓘ Only those human *activities* (such as construction work, mining or drilling) that could result in direct disturbance of the *disposal facility* (i.e. disturbance of the *waste* itself, of the contaminated *near field* or of materials of the engineered *barrier*) are included.

human-machine interface

The part of a system through which personnel interact with the system to perform their functions and tasks.

ⓘ The human–machine interface constitutes the interface between personnel and plant systems, including procedures, communication systems displays, *alarms* and controls.

human motor control

The physiological capability of a human's muscular system to control movement, including strength and fine movements.

hydrodynamic dispersion

See *dispersion*.

hypocentre

The point (focus) within the Earth at which an earthquake is initiated.

[hypothetical critical group]

See [*critical group*].

I

IAEA publication

An IAEA copyrighted hard copy or electronic product issued with unlimited distribution and bearing the IAEA emblem (logo) on the front and officially approved by the Publications Committee on behalf of the Director General.

> ⓘ An IAEA document is an official non-copyrighted hard copy or electronic product issued with limited distribution and bearing the IAEA emblem (logo) on the front.

> ⓘ A manuscript is an unissued copy of a draft publication or a draft document.

> ⓘ A TECDOC is a publication, not a document.

ICRU sphere

A sphere of 30 cm diameter made of *tissue equivalent material* with a density of 1 g/cm^3 and a mass composition of 76.2% oxygen, 11.1% carbon, 10.1% hydrogen and 2.6% nitrogen.

> ⓘ The *ICRU (International Commission on Radiation Units and Measurements) sphere* is used as a reference phantom in defining *dose equivalent quantities*.

See Ref. [41].

igneous rock

Rock that has formed from *magma*.

> ⓘ Extruded *igneous rocks* (volcanic rocks) are typically divided into four basic types according to their SiO_2 content: basalt, andesite, dacite and rhyolite.

ignition source

An applied (external) source of heat which is used to ignite *combustible materials*.

immediate dismantling

See *decommissioning* (1).

immobilization

See *radioactive waste management* (1).

important human task

A human task that can have an adverse or positive effect on *safety*, as determined by *safety analysis*.

in-service inspection

See *inspection*.

in-structure response spectrum

See *response spectrum.*

incident

Any unintended *event*, including operating errors, equipment *failures, initiating events, accident precursors, near misses* or other mishaps, or unauthorized act, malicious or non-malicious, the consequences or potential consequences of which are not negligible from the point of view of *protection and safety.*

> See also *event* and *International Nuclear and Radiological Event Scale.*

> ! The word *incident* is sometimes used, for example in the INES 2008 Manual [16], to describe *events* that are, in effect, minor *accidents*, i.e. that are distinguished from *accidents* only in terms of being less severe.

> ! This is a distinction with little basis in general usage, in which an *incident* can be minor or major, just as an *accident* can; however, unlike an *accident*, an *incident* can be caused intentionally.

> ⓘ The definition of *incident* given was derived on the basis of the entries for *accident* and *event* and the explanation of the term *incident* given in SF-1 [24].

> **[*nuclear incident*].** Any occurrence or series of occurrences having the same origin which causes *nuclear damage* or, but only with respect to preventive measures, creates a grave and imminent threat of causing such damage. (See Ref. [42].)

> ! This usage is specific to the Convention on Supplementary Compensation for Nuclear Damage [42], for the purposes of the Convention, and should otherwise be avoided.

See also [*nuclear damage*].

incident commander

See *emergency response commander.*

independent assessment

See *assessment* (2).

independent equipment

Equipment that possesses both of the following characteristics:

(a)   The ability to perform its required function is unaffected by the *operation* or *failure* of other equipment.

(b)   The ability to perform its required function is unaffected by the occurrence of the effects resulting from the *initiating event* for which it is required to function.

indicator

> **condition indicator.** Characteristic of a *structure, system or component* that can be observed, measured or trended to infer or directly indicate the current and future ability of the *structure, system or component* to function within *acceptance criteria.*

functional indicator. Condition indicator that is a direct indication of the current ability of a *structure, system or component* to function within *acceptance criteria.*

performance indicator. Characteristic of a *process* that can be observed, measured or trended to infer or directly indicate the current and future performance of the *process*, with particular emphasis on satisfactory performance for *safety.*

individual dose

See *dose concepts.*

individual monitoring

See *monitoring* (1).

individualization

The ability to associate a forensic result or a set of results uniquely with a single source, such as a person, place or production process.

INES

See *International Nuclear and Radiological Event Scale (INES).*

infant

ⓘ In dosimetry, unless otherwise stated, an *infant* is assumed to be three months old, and annual quantities (e.g. *annual dose*, annual *intake*) relating to an *infant* refer to the year starting at birth.

ⓘ The values for the three month old *infant* are intended to be valid for the first year of life.

ⓘ In common usage for internal dosimetry an *infant* is taken to be 100 days old.

See also *child* and *reference individual.*

information alert

Time sensitive reporting that could indicate a *nuclear security event*, requiring assessment, and may come from a variety of sources, including operational information, medical surveillance, accounting and *consigner*/consignee discrepancies, border monitoring, etc.

information object

Knowledge or data that have value to the organization.

information security

The preservation of the *confidentiality*, *integrity* and *availability* of information.

'informed customer' capability

ⓘ The capability of an organization to have a clear knowledge and understanding of the product being supplied or the service being provided.

ingestion and commodities planning distance (ICPD)

See *emergency planning distance*.

[inhalation class]

See *lung absorption type*.

initial entry

The first controlled entry made into a *crime scene*, conducted for the purpose of gathering data regarding the nature and extent of on-scene hazards.

initiating event

An identified *event* that leads to *anticipated operational occurrences* or *accident conditions*.

ⓘ This term (often shortened to **initiator**) is used in relation to *event* reporting and *analysis*; that is, when such *events* have occurred.

ⓘ For the consideration of hypothetical *events* at the *design* stage, the term *postulated initiating event* is used.

postulated initiating event (PIE). A postulated *event* identified in *design* as capable of leading to *anticipated operational occurrences* or *accident conditions*.

ⓘ The primary causes of *postulated initiating events* may be credible equipment *failures* and *operator* errors (both within and external to the *facility*), human induced *events* or natural *events*.

initiator

See *initiating event*.

inner cordoned off area

An area established by *first responders* in an *emergency* around a potential *radiation hazard*, within which *protective actions* and *other response actions* are taken to protect *first responders* and the public from possible *exposure* and *contamination*.

innocent alarm

See *alarm*.

insider

An individual with authorized access to *associated facilities* or *associated activities* or to *sensitive information* or *sensitive information assets*, who could commit, or facilitate the commission of criminal or intentional unauthorized acts involving or directed at *nuclear material*, *other radioactive material*, *associated facilities* or *associated activities* or other acts determined by the State to have an adverse impact on nuclear security.

! In general, an *insider* is somebody with the relevant access, but not necessarily any motivation or intent to commit *malicious acts*.

ⓘ In some Nuclear Security Series publications, this term has also been defined in a broadly similar way as "One or more individuals with authorized access to *nuclear facilities* or *nuclear material* in *transport* who could attempt *unauthorized removal* or *sabotage*, or who could aid an *external adversary* to do so" [6] and "An individual with authorized access to *associated facilities* or *associated activities* or to *sensitive information* or *sensitive information assets*, who could commit, or facilitate the commission of a *malicious act*" [7].

insider adversary. See *adversary*.

unwitting insider. An *insider* without the intent and motivation to commit a *malicious act* who is exploited by an *adversary* without the unwitting insider's awareness.

See also *adversary*.

insider adversary

See *adversary*.

inspection

1.	An *examination*, observation, *surveillance*, measurement or test undertaken to assess *structures, systems and components* and materials, as well as operational *activities*, technical *processes*, organizational *processes*, *procedures* and personnel competence.

> **in-service inspection.** *Inspection* of *structures, systems and components* undertaken over the *operating lifetime* by or on behalf of the *operating organization* for the purpose of identifying age related degradation or conditions that, if not addressed, might lead to the *failure* of *structures, systems or components*.

> ⓘ *Inspection* of operational *activities, processes*, etc., by or on behalf of the *operating organization* would normally be described by terms such as *self-assessment* and *audit*.

> **regulatory inspection.** *Inspection* undertaken by or on behalf of the *regulatory body*.

2.	An evaluation of the conformity to a *requirement*.

inspection imaging device

See *device*.

institutional control

See *control* (1).

instrument alarm

See *alarm*.

intake

1.	The act or *process* of taking radionuclides into the body by inhalation or ingestion or through the skin.

> ⓘ Other *exposure pathways* by *intake* are injection (e.g. in nuclear medicine) and *intake* via a wound, as distinguished from *intake* through (intact) skin.

2.	The *activity* of a radionuclide taken into the body in a given time period or as a result of a given *event*.

104

acute intake. An *intake* occurring within a time period short enough that it can be treated as instantaneous for the purposes of assessing the resulting *committed dose*.

! The *exposure* that results from an *acute intake* is not necessarily *acute exposure*. For a long lived radionuclide that is retained in the body, an *acute intake* will result in chronic (i.e. long term) *exposure*.

chronic intake. An *intake* over an extended period of time, such that it cannot be treated as a single instantaneous *intake* for the purposes of assessing the resulting *committed dose*.

ⓘ *Chronic intake* may, however, be treated as a series of *acute intakes*.

integrated management system

See management system.

integrity

The property of accuracy and completeness of information.

interacting event

An *event* or a sequence of associated *events* that, interacting with a *facility*, affect *site personnel* or *items important to safety* in a manner that could adversely influence *safety*.

interested party

A person, company, etc., with a concern or interest in the activities and performance of an organization, business, system, etc.

ⓘ The term *interested party* is used in a broad sense to mean a person or group having an interest in the performance of an organization.

ⓘ Those who can influence events may effectively become *interested parties* — whether their 'interest' is regarded as 'genuine' or not — in the sense that their views need to be considered.

ⓘ *Interested parties* would need to be specified as relevant.

ⓘ *Interested parties* have typically included the following: customers, owners, *operators*, employees, *suppliers*, partners and trade unions; the regulated industry or professionals; scientific bodies; governmental agencies or *regulatory bodies* (national, regional and local) whose responsibilities may cover nuclear energy; the media; the public (individuals, community groups and interest groups); and other States, especially neighbouring States that have entered into agreements providing for an exchange of information concerning possible transboundary impacts, or States involved in the export or import of certain technologies or materials [43].

! The term [***stakeholder***] is used in the same broad sense as *interested party* and the same provisos are necessary.

! The term *stakeholder* has disputed usage, and it is misleading and too all-encompassing for clear use. In view of the potential for misunderstanding and misrepresentation, use of the term is discouraged in favour of *interested party*.

ⓘ The Handbook on Nuclear Law [43] states that: "Owing to the differing views on who has a genuine interest in a particular nuclear related activity, no authoritative definition of stakeholder has yet been offered, and no definition is likely to be accepted by all parties."

[interim storage]

See *storage*.

intermediate bulk container (IBC)

A portable *packaging* that:

(a)   Has a capacity of not more than 3 m³;

(b)   Is designed for mechanical handling;

(c)   Is resistant to the stresses produced in handling and *transport*, as determined by tests.

(See SSR-6 (Rev. 1) [2].)

intermediate level waste (ILW)

See *waste classes*.

internal exposure

See *exposure* (1).

internal hazard

See hazard.

International Nuclear and Radiological Event Scale (INES)

ⓘ The *INES* is a scale developed for use by States for the purpose of communicating with the public on the safety significance of *events* associated with *sources* of *radiation*.

! The *INES* should not be confused with the *emergency classification* system, and should not be used as a basis for *emergency response actions*.

! In the 2008 INES Manual [16], there was a fundamental mismatch between the terminology and usage in *safety standards* and the designations used in *INES*.

! The *INES* 2008 terminology — in particular the use of the terms *incident* and *accident* — was different from that in *safety standards* and from the usual English meanings of the words, and great care should be taken to avoid confusion between the two areas.

! In short, *events* that would be considered *accidents* according to the *safety standards* definition may be *accidents* or *incidents* (i.e. not *accidents*) in *INES* 2008 terminology.

ⓘ This was not a serious day to day problem because the two areas are quite separate and have quite different purposes. However, it was a potential cause of confusion in communication with the news media and the public.

ⓘ See the information notes associated with the terms *event*, *incident* and *accident* for further information.

[international nuclear transport]

See *transport* (1).

interplate tectonic processes

Tectonic processes occurring at the interfaces between the Earth's tectonic plates.

intervention

Any action intended to reduce or avert *exposure* or the likelihood of *exposure* due to *sources* that are not part of a controlled *practice* or that are out of *control* as a consequence of an *accident*.

ⓘ This definition is somewhat more explicit than (though not necessarily inconsistent with) that of Ref. [44].

ⓘ The term *facilities and activities* is intended to provide an alternative to the terminology of *sources* and *practices* (or *interventions*) to refer to general categories of situations.

ⓘ In *emergency preparedness* and *response*, the concepts of *protective actions* and protection strategy are now used instead.

intraplate

Of tectonic processes, within the Earth's tectonic plates.

intrusion (human)

See *human intrusion*.

intrusion barrier

See *barrier*.

inventory

book inventory. The algebraic sum of the previous *physical inventory* (as determined at a physical inventory taking) and any subsequent inventory changes (as reflected in the inventory change reports).

physical inventory. The sum of all the measured or derived estimates of batch quantities of *nuclear material* physically present at a given time within a *material balance area*, obtained by a facility operator in accordance with specified procedures.

§ These terms are used in nuclear security, The term *inventory* is also widely used in safety, especially in radioactive waste safety, to refer to the total amount of *radioactive material, radioactive sources* or *radioactive waste* within a certain specified area or intended to be managed in a certain specified way (or a breakdown of the characteristics of the material or waste within that total amount, for example the total activity of each radionuclide present).

investigation level

See *level*.

iodine thyroid blocking

The administration of a compound of stable iodine (usually potassium iodide) to prevent or reduce the *uptake* of *radioactive* isotopes of iodine by the thyroid in a *nuclear or radiological emergency* involving *radioactive* iodine.

(i) *Iodine thyroid blocking* is an *urgent protective action*.

(i) The terms 'stable iodine prophylaxis', 'thyroid blocking' or 'iodine blockade' are sometimes used to describe the same concept, but *iodine thyroid blocking* is preferred in *IAEA publications*.

ionizing radiation

See *radiation*.

irradiation installation

A *structure* or an installation that houses a particle accelerator, X ray apparatus or large *radioactive source* and that can produce high *radiation* fields.

(i) *Irradiation installations* include installations for external beam *radiation* therapy, installations for sterilization or preservation of commercial products and some installations for industrial radiography.

irregularity

An unusual observable condition which might result from *unauthorized removal* of *nuclear material*, or which restricts the ability of the facility operator to draw the conclusion that *unauthorized removal* has not occurred.

isolation (of radioactive waste in a disposal facility)

The physical separation and retention of *radioactive waste* away from people and from the *environment*.

(i) *Isolation* of *radioactive waste* with its associated *hazards* in a *disposal facility* involves the minimization of the influence of factors that could reduce the integrity of the *disposal facility*; provision for a very low mobility of most long lived radionuclides to impede their *migration* from the *disposal facility*; and making access to the waste by people difficult without special technical capabilities.

(i) Design features are intended to provide *isolation* (a *confinement* function) for several hundreds of years for *short lived waste* and for at least several thousand years for *intermediate level waste* and *high level waste*. *Isolation* is an inherent feature of *geological disposal*.

IT security

See *computer security*.

item important to safety

See *plant equipment (for a nuclear power plant)*.

J

justification

1.   The *process* of determining for a *planned exposure situation* whether a *practice* is, overall, beneficial; that is, whether the expected benefits to individuals and to society from introducing or continuing the *practice* outweigh the harm (including *radiation detriment*) resulting from the *practice*.

2.   The *process* of determining for an *emergency exposure situation* or an *existing exposure situation* whether a proposed *protective action* or *remedial action* is likely, overall, to be beneficial; that is, whether the expected benefits to individuals and to society (including the reduction in *radiation detriment*) from introducing or continuing the *protective action* or *remedial action* outweigh the cost of such action and any harm or damage caused by the action.

K

kerma, *K*

The quantity *K*, defined as:

$$K = \frac{dE_{tr}}{dm}$$

where dE_{tr} is the sum of the initial kinetic energies of all charged ionizing particles liberated by uncharged ionizing particles in a material of mass d*m*.

ⓘ The SI unit for *kerma* is joules per kilogram (J/kg), termed the *gray* (Gy).

ⓘ *Kerma* was originally an acronym for kinetic energy released in matter but is now accepted as a word.

air kerma. The *kerma* value for air.

ⓘ Under charged particle equilibrium conditions, the *air kerma* (in *gray*) is numerically approximately equal to the *absorbed dose* in air (in *gray*).

reference air kerma rate. The *kerma* rate to air, in air, at a reference distance of 1 m, corrected for air *attenuation* and scattering.

ⓘ This quantity is expressed in μGy/h at 1 m.

kerma factor

The *kerma* per unit *particle fluence*.

knowledge management

An integrated, systematic approach to identifying, managing and sharing an organization's knowledge and enabling groups of people to create new knowledge collectively to help in achieving the organization's objectives.

ⓘ In the context of *management systems*, *knowledge management* helps an organization to gain insight and understanding from its own experience.

ⓘ Specific *activities* in *knowledge management* help the organization to better acquire, record, store and utilize knowledge.

ⓘ The term 'knowledge' is often used to refer to bodies of facts and principles accumulated by humankind over the course of time.

ⓘ Explicit knowledge is knowledge that is contained in, for example, documents, drawings, calculations, designs, databases, procedures and manuals.

ⓘ Tacit knowledge is knowledge that is held in a person's mind and has typically not been captured or transferred in any form (if it were, it would become explicit knowledge).

ⓘ Knowledge is distinct from information: data yield information and knowledge is gained by acquiring, understanding and interpreting information.

ⓘ Knowledge and information each consist of true statements, but knowledge serves a purpose: knowledge confers a capacity for effective action.

ⓘ Knowledge for an organization is the acquiring, understanding and interpreting of information.

ⓘ Knowledge may be applied for such purposes as: problem solving and learning; forming judgements and opinions; decision making, forecasting and strategic planning; generating feasible options for action and taking actions to achieve desired results.

ⓘ Knowledge also protects intellectual assets from decay, augments intelligence and provides increased flexibility.

L

large freight container

See *freight container*.

large release of radioactive material

A *release* of *radioactive material* for which *off-site protective actions* that are limited in terms of times and *areas* of application are insufficient for protecting people and the *environment*.

    ⓘ See also *early release of radioactive material*; see also *defence in depth (1)*.

late effect

See *health effects (of radiation)*.

latent weakness

See *cause*.

lava

Molten rock erupted at the Earth's surface by a *volcano* or by an eruptive fissure as an effusive dome or flow.

    ⓘ When first emitted from a *volcanic vent*, *lava* is a liquid at very high temperature, typically 700–1200°C.

    ⓘ *Lava* flows vary by many orders of magnitude in their viscosities and this strongly influences their flow properties.

[legal person]

Any organization, corporation, partnership, firm, association, trust, estate, public or private institution, group, political or administrative entity or other person designated in accordance with national legislation who or which has responsibility and authority for any action having implications for *protection and safety*.

    ⓘ Contrasted in legal texts with 'natural person', meaning an individual.

    ⓘ Superseded by the term *person or organization*, which should be used.

See also *applicant*, *licence* and *registration*.

level

    clearance level. A value, established by a *regulatory body* and expressed in terms of *activity concentration*, at or below which *regulatory control* may be removed from a *source* of *radiation* within a notified or authorized *practice*.

    See also *clearance (1)*.

    diagnostic reference level. A level used in medical imaging to indicate whether, in routine conditions, the *dose* to the *patient* or the amount of radiopharmaceuticals administered in a specified radiological procedure for medical imaging is unusually high or unusually low for that procedure [1].

ⓘ For the use of radiopharmaceuticals, the *diagnostic reference level* is a level of *activity* for typical *examinations* for groups of standardized *patients* or for a standard phantom and for broadly defined types of equipment.

ⓘ The *diagnostic reference levels* are indicative of good practice, when not exceeded, for standard procedures in which good practices and normal practices are applied with regard to diagnostic performance and technical performance.

emergency action level (EAL). A specific, predetermined criterion for observable conditions used to detect, recognize and determine the *emergency class*.

ⓘ An *emergency action level* could represent an instrument reading, the status of a piece of equipment or any observable *event*, such as a fire.

exemption level. A value, established by a *regulatory body* and expressed in terms of *activity concentration*, total *activity*, *dose rate* or *radiation* energy, at or below which a *source* of *radiation* need not be subject to some or all aspects of *regulatory control*.

ⓘ A *regulatory body* may also grant *exemption* on a case by case basis, following *notification*.

ⓘ Although the term *exemption level* does not strictly apply in such a situation, a criterion for *exemption* may nevertheless be established by the *regulatory body*, expressed in similar terms or, alternatively, expressed in terms of *annual dose* on the basis of an appropriate *dose assessment*. (See GSR Part 3 [1] and para. 5.12 of RS-G-1.7 [20].)

ⓘ Values of *exemption levels* are specified in table I.1 and table I.2 of schedule I of GSR Part 3 [1].

investigation level. The value of a quantity such as *effective dose*, *intake* or *contamination* per unit area or volume at or above which an investigation would be conducted.

operational intervention level (OIL). A set *level* of a measurable quantity that corresponds to a generic criterion.

ⓘ *Operational intervention levels* are typically expressed in terms of *dose rates* or of *activity* of *radioactive material* released, time integrated air *activity concentrations*, ground or surface concentrations, or *activity concentrations* of radionuclides in environmental, *food* or water samples.

ⓘ *An operational intervention level* is used immediately and directly (without further *assessment*) to determine the appropriate *protective actions* on the basis of an environmental measurement.

recording level. A level of *dose, exposure* or *intake* specified by the *regulatory body* at or above which values of *dose* to, *exposure* of or *intake* by *workers* are to be entered in their individual *exposure* records.

reference level. For an *emergency exposure situation* or an *existing exposure situation*, the level of *dose, risk* or *activity concentration* above which it is not appropriate to plan to allow *exposures* to occur and below which *optimization of protection and safety* would continue to be implemented.

ⓘ The value chosen for a *reference level* will depend upon the prevailing circumstances for the *exposure* under consideration.

licence

1.   A legal document issued by the *regulatory body* granting *authorization* to perform specified *activities* relating to a *facility or activity*.

> ⓘ A *licence* is a product of the *authorization process* (although the term **licensing process** is sometimes used), and a *practice* with a current *licence* is an authorized *practice*.

> ⓘ *Authorization* may take other forms, such as *registration* or *certification*.

2.   [Any *authorization* granted by the *regulatory body* to the *applicant* to have the responsibility for the *siting, design, construction, commissioning, operation* or *decommissioning* of a *nuclear installation*.] (See Ref. [10].)

3.   [Any *authorization*, permission or *certification* granted by a *regulatory body* to carry out any *activity* related to management of *spent fuel* or of *radioactive waste*.] (See Ref. [11].)

> ! The definitions (2) and (3) from the Conventions [10, 11] are somewhat more general in scope than the usual IAEA usage in definition (1).

> ! In IAEA usage, a *licence* is a particular type of *authorization*, normally representing the primary *authorization* for the *operation* of a whole *facility* or *activity*.

> ⓘ The conditions attached to the *licence* may require that further, more specific, *authorization* or *approval* be obtained by the *licensee* before carrying out particular *activities*.

licensee

> ⓘ The holder of a current *licence*. The *licensee* is the *person or organization* having overall responsibility for a *facility* or *activity*.

licensing basis

A set of regulatory *requirements* applicable to a *nuclear installation*.

> ⓘ The *licensing basis*, in addition to a set of regulatory *requirements*, may also include agreements and commitments made between the *regulatory body* and the *licensee* (e.g. in the form of letters exchanged or of statements made in technical meetings).

licensing process

See *licence* (1).

life, lifetime

design life. The period of time during which a *facility* or *component* is expected to perform according to the technical specifications to which it was produced.

operating lifetime, operating life

1. The period during which an *authorized facility* is used for its intended purpose, until *decommissioning* or *closure*.

> ⓘ The synonyms **operating period** and **operational period** are also used.

2. [The period during which a *spent fuel* or a *radioactive waste* management *facility* is used for its intended purpose. In the case of a *disposal facility*, the period begins when *spent fuel* or *radioactive waste* is first emplaced in the *facility* and ends upon *closure* of the *facility*.] (See Ref. [11].)

qualified life. Period for which a *structure, system or component* has been demonstrated, through testing, *analysis* or experience, to be capable of functioning within *acceptance criteria* during specific *operating conditions* while retaining the ability to perform its *safety functions* in *accident conditions* for a *design basis accident* or a *design basis* earthquake.

See also *equipment qualification, specified service conditions, harsh environment, mild environment.*

service life. The period from initial *operation* to final withdrawal from service of a *structure, system or component.*

life cycle management

Life management (or *lifetime management*) in which due recognition is given to the fact that at all stages in the *lifetime* there may be effects that need to be taken into consideration.

ⓘ An example is the approach to products, *processes* and services in which it is recognized that at all stages in the *lifetime* of a product (extraction and processing of raw materials, manufacturing, *transport* and distribution, use and *reuse*, and *recycling* and *waste* management) there are environmental impacts and economic consequences.

ⓘ The term 'life cycle' (as opposed to *lifetime*) implies that the life is genuinely cyclical (as in the case of *recycling* or *reprocessing*).

See also *'cradle to grave' approach* and *ageing management.*

life management

See *ageing management.*

lifetime

See *life, lifetime.*

lifetime dose

See *dose concepts.*

lifetime management

See *ageing management.*

lifetime risk

See *risk* (3).

limit

The value of a quantity used in certain specified *activities* or circumstances that must not be exceeded.

! The term *limit* should only be used for a criterion that must not be exceeded; for example, where exceeding the *limit* would cause some form of legal sanction to be invoked.

! Criteria used for other purposes — for example, to indicate a need for closer investigation or a review of *procedures*, or as a threshold for reporting to a *regulatory body* — should be described using other terms, such as *reference level*.

acceptable limit. A *limit* acceptable to the *regulatory body*.

ⓘ The term *acceptable limit* is usually used to refer to a *limit* on the predicted radiological consequences of an *accident* (or on *potential exposures* if they occur) that is acceptable to the relevant *regulatory body* when the probability of occurrence of the *accident* or *potential exposures* has been taken into account (i.e. on the basis that it is unlikely to occur).

ⓘ The term *authorized limit* should be used to refer to *limits* on *doses* or *risks*, or on *releases* of radionuclides, which are acceptable to the *regulatory body* on the assumption that they are likely to occur.

annual limit on exposure (ALE). The *potential alpha energy exposure* in a year that would result in inhalation of the *annual limit on intake (ALI)*.

ⓘ Used for *exposure* due to decay products of ^{222}Rn or ^{220}Rn.

ⓘ In units of J·h/m^3.

annual limit on intake (ALI). The *intake* by inhalation or ingestion or through the skin of a given radionuclide in a year by the *reference individual* which would result in a *committed dose* equal to the relevant *dose limit*.

ⓘ The *annual limit on intake* is expressed in units of *activity*.

See Refs [28, 29].

authorized limit. A *limit* on a measurable quantity, established or formally accepted by a *regulatory body*.

! Wherever possible, *authorized limit* should be used in preference to *prescribed limit*.

ⓘ Equivalent in meaning to *prescribed limit*, *authorized limit* has been more commonly used in *radiation safety* and the *safety* of *radioactive waste management*, in particular in the context of *limits* on *discharges*.

derived limit. A *limit* on a measurable quantity set, on the basis of a *model*, such that compliance with the *derived limit* may be assumed to ensure compliance with a *primary limit*.

dose limit. The value of the *effective dose* or the *equivalent dose* to individuals in *planned exposure situations* that is not to be exceeded.

operational limits and conditions. A set of rules setting forth parameter *limits*, the functional capability and the performance levels of equipment and personnel approved by the *regulatory body* for safe *operation* of an *authorized facility*.

[prescribed limit]. A *limit* established or accepted by the *regulatory body*.

ⓘ The term *authorized limit* is preferred.

primary limit. A *limit* on the *dose* or *risk* to an individual.

safety limits. *Limits* on operational parameters within which an *authorized facility* has been shown to be safe.

ⓘ *Safety limits* are beyond the *limits* for *normal operation*.

116

[*secondary limit*]. A *limit* on a measurable quantity that corresponds to a *primary limit*.

> ! Such a *limit* meets the definition of *derived limit*, and *derived limit* should be used.

> ⓘ For example, the *annual limit on intake*, a *derived limit*, corresponds to the *primary limit* on annual *effective dose* for a *worker*.

limited access area

See *area*.

linear energy transfer (LET), L_Δ

Defined generally as:

$$L_\Delta = \left(\frac{\mathrm{d}E}{\mathrm{d}\ell}\right)_\Delta$$

where $\mathrm{d}E$ is the energy lost in traversing distance $\mathrm{d}\ell$ and Δ is an upper bound on the energy transferred in any single collision.

> ⓘ A measure of how, as a function of distance, energy is transferred from *radiation* to the exposed matter. A high value of *linear energy transfer* indicates that energy is deposited within a small distance.

> ⓘ L_∞ (i.e. with $\Delta = \infty$) is termed the **unrestricted linear energy transfer** in defining the *quality factor*.

> ⓘ L_Δ is also known as the **restricted linear collision stopping power**.

linear–no threshold (LNT) hypothesis

The hypothesis that the *risk* of *stochastic effects* is directly proportional to the *dose* for all levels of *dose* and *dose rate* below those levels at which *deterministic effects* occur.

> ⓘ That is, that any non-zero *dose* implies a non-zero *risk* of *stochastic effects*.

> ⓘ This is the working hypothesis on which the IAEA's *safety standards* (and the International Commission on Radiological Protection's recommendations) are based.

> ⓘ The hypothesis is not proven — indeed it is probably not provable — for low *doses* and *dose rates*, but it is considered the most defensible assumption in radiobiological terms on which to base *safety standards*.

> ⓘ Other hypotheses conjecture that the *risk* of *stochastic effects* at low *doses* and/or *dose rates* is:

> > (a) Greater than that implied by the *linear–no threshold hypothesis* (superlinear hypotheses);

> > (b) Less than that implied by the *linear–no threshold hypothesis* (sublinear hypotheses);

> > (c) Zero below some threshold value of *dose* or *dose rate* (threshold hypotheses); or

> > (d) Negative below some threshold value of *dose* or *dose rate*, that is, that low *doses* and *dose rates* protect individuals against *stochastic effects* and/or other types of harm (hormesis hypotheses).

'living' probabilistic safety assessment

See *probabilistic safety assessment (PSA)*.

logic

The generation of a required binary output signal from a number of binary input signals according to predetermined rules.

ⓘ The term is also applied to the types of equipment used for generating this signal (e.g. *logic* gate, *logic* board).

long lived waste

See *waste classes*.

low dispersible radioactive material

Either solid *radioactive material*, or solid *radioactive material* in a sealed capsule, that has limited dispersibility and is not in powder form. (See SSR-6 (Rev. 1) [2].)

! This usage is specific to the Transport Regulations [2], and should otherwise be avoided.

low enriched uranium (LEU)

See *uranium*.

low level waste (LLW)

See *waste classes*.

low linear energy transfer (LET) radiation

See *radiation*.

low specific activity (LSA) material

Radioactive material that by its nature has a limited *specific activity*, or *radioactive material* for which *limits* of estimated average *specific activity* apply. (See SSR-6 (Rev. 1) [2].)

! External shielding materials surrounding the *low specific activity material* are required not to be considered in determining the estimated average *specific activity*.

! This usage is specific to the Transport Regulations [2], and should otherwise be avoided.

low toxicity alpha emitters

Natural uranium; *depleted uranium*; natural thorium; ^{235}U or ^{238}U; ^{232}Th; ^{228}Th and ^{230}Th when contained in ores, or in physical and chemical concentrates; or alpha emitters with a *half-life* of less than 10 days. (See SSR-6 (Rev. 1) [2].)

lower limit of detection

See *minimum detectable activity (MDA)*.

lung absorption type

A classification used to distinguish between the different rates at which inhaled radionuclides are transferred from the respiratory tract to the blood.

ⓘ Reference [45] classifies materials into four *lung absorption types*:

(a) Type V (very fast) are materials that, for dosimetric purposes, are assumed to be instantaneously absorbed into the blood;

(b) Type F (fast) are materials that are readily absorbed into the blood;

(c) Type M (moderate) are materials that have intermediate rates of absorption into the blood;

(d) Type S (slow) are materials that are relatively insoluble and are only slowly absorbed into the blood.

ⓘ The *lung absorption types* supersede the **[inhalation classes]** D (days), M (months) and Y (years) previously recommended in Refs [27–29] (often referred to informally as 'lung classes').

ⓘ There is an approximate correspondence between *lung absorption type* F and *inhalation class* D, between *lung absorption type* M and *inhalation class* M and between *lung absorption type* S and *inhalation class* Y.

See also *gut transfer factor*, a similar concept for ingested radionuclides in the gastrointestinal tract.

M

magma

A mixture of molten rock (800–1200°C) which can also contain suspended crystals, dissolved gases and sometimes gas bubbles.

    ⓘ *Magma* forms by the melting of existing rock in the *Earth's crust* or in the *Earth's mantle*.

    ⓘ *Magma* composition and gas content generally control the type of *eruption* at a *volcano*.

    ⓘ In general terms, hotter, less viscous *magma* (e.g. basalt) allows gas to separate more efficiently, limiting the explosivity of the *eruption*, while cooler, more viscous *magma* (e.g. andesite, dacite, rhyolite) is more likely to fragment violently during *eruption*.

magma chamber

An underground reservoir that is filled with *magma* and tapped during a *volcanic eruption*.

    ⓘ *Magma* in these reservoirs can partially crystallize or mix with new *magma*, which can change the *eruption* composition or *hazard* over time.

magnitude (of an earthquake)

Measure of the size of an earthquake relating to the energy released in the form of seismic waves.

    ⓘ Seismic *magnitude* means the numerical value on a standardized scale such as, but not limited to, moment *magnitude*, surface wave *magnitude*, body wave *magnitude*, local *magnitude* or duration *magnitude*.

    maximum potential magnitude. Reference value used in seismic *hazard analysis* characterizing the potential of a seismic source to generate earthquakes.

    ⓘ The way in which the *maximum potential magnitude* is calculated depends on the type of seismic source considered and the approach to be used in the seismic *hazard* analysis.

See also beyond design basis earthquake.

main safety function

See *safety function*.

maintenance

The organized *activity*, both administrative and technical, of keeping *structures, systems and components* in good operating condition, including both preventive and corrective (or *repair*) aspects.

    corrective maintenance. Actions that restore, by *repair*, overhaul or replacement, the capability of a failed *structure, system or component* to function within *acceptance criteria*.

    ⓘ *Corrective maintenance* does not necessarily result in a significant extension of the expected useful *life* of a functional *structure, system or component*.

    ⓘ Contrasted with *preventive maintenance*.

periodic maintenance. Form of *preventive maintenance* consisting of servicing, parts replacement, *surveillance* or testing at predetermined intervals of calendar time, operating time or number of cycles.

ⓘ Also termed *time based maintenance*.

planned maintenance. Form of *preventive maintenance* consisting of refurbishment or replacement that is scheduled and performed prior to unacceptable degradation of a *structure, system or component*.

predictive maintenance. Form of *preventive maintenance* performed continuously or at intervals governed by observed condition to monitor, diagnose or trend *condition indicators* of a *structure, system or component*; results indicate present and future functional ability or the nature of and schedule for *planned maintenance*.

ⓘ Also termed *condition based maintenance*.

preventive maintenance. Actions that detect, preclude or mitigate degradation of a functional *structure, system or component* to sustain or extend its useful *life* by controlling degradation and *failures* to an acceptable level.

ⓘ *Preventive maintenance* may be *periodic maintenance, planned maintenance* or *predictive maintenance*.

ⓘ Contrasted with *corrective maintenance*.

reliability centred maintenance. A *process* for specifying applicable *preventive maintenance requirements* for *items important to safety* and equipment in order to prevent potential *failures* or to control the *failure modes* optimally.

ⓘ *Reliability centred maintenance* utilizes a decision *logic* tree to identify the *maintenance requirements* according to the *safety* consequences and operational consequences of each *failure* and the degradation mechanism responsible for the *failures*.

maintenance bypass

See *bypass* (1).

major public event

A high-profile event that a State has determined to be a potential *target*.

ⓘ Major public events include sporting, political and religious gatherings involving large numbers of spectators and participants.

! This refers to a different type of 'event' from a *nuclear security event*.

ⓘ In publications not specifically addressing security for major public events, it should not normally be necessary to give a definition.

malicious act

An act or attempt of unauthorized removal of radioactive material or sabotage.

ⓘ A '*criminal act*' is normally covered by criminal or penal law in a State, whereas an '*unauthorized act*' is typically the subject of administrative or civil law. In addition, criminal acts involving nuclear or other radioactive material may constitute offences related to acts of terrorism, including those set out in the Convention on the Physical Protection of Nuclear Material and its Amendment and the International Convention for the

121

Suppression of Acts of Nuclear Terrorism, all of which, in some States, are subject to special legislation. Unauthorized acts with nuclear security implications could include both intentional and unintentional unauthorized acts as determined by the State. Examples of a criminal act or an unauthorized act with nuclear security implications could, if determined by the State, include: (1) the undertaking of an unauthorized activity involving radioactive material by an authorized person; (2) the unauthorized possession of radioactive material by a person with the intent to commit a criminal or unauthorized act with such material, or to facilitate the commission of such acts; or (3) the failure of an authorized person to maintain adequate control of radioactive material, thereby making it accessible to persons intending to commit a criminal or an unauthorized act, using such material.

management (of sealed radioactive sources)

[The administrative and operational activities that are involved in the manufacture, supply, receipt, possession, *storage*, use, transfer, import, export, transport, maintenance, *recycling* or *disposal* of *radioactive sources*.] (See Ref. [21].)

> ! This usage is specific to the Code of Conduct on the Safety and Security of Radioactive Sources [21].

management system

A set of interrelated or interacting elements (*system*) for establishing policies and objectives and enabling the objectives to be achieved in an efficient and effective manner.

> ⓘ The component parts of the *management system* include the organizational structure, resources and organizational *processes*.

> ⓘ Management is defined (in ISO 9000) [46] as coordinated *activities* to direct and *control* an organization.

> ⓘ The *management system* integrates all elements of an organization into one coherent system to enable all of the organization's objectives to be achieved. These elements include the organizational structure, resources and *processes*.

> ⓘ Personnel, equipment and organizational culture as well as the documented policies and *processes* form parts of the *management system*.

> ⓘ The organization's *processes* have to address the totality of the *requirements* on the organization as established in, for example, IAEA *safety standards* and other international codes and standards.

> ***integrated management system.*** A single coherent *management system* for *facilities and activities* in which all the component parts of an organization are integrated to enable the organization's objectives to be achieved.

> ⓘ These component parts of an organization that are integrated include the organizational structure, resources and organizational *processes*.

management system review

A regular and systematic evaluation by *senior management* of an organization of the suitability, adequacy, effectiveness and efficiency of its *management system* in executing the policies and achieving the goals and objectives of the organization.

mantle, Earth's

See *Earth's mantle*.

material ageing

See *ageing: physical ageing*.

material balance area

An area in a *nuclear facility* designated such that: (a) the quantity of *nuclear material* in each movement into or out of each material balance area can be determined; and (b) the *physical inventory* of *nuclear material* in each material balance area can be determined when necessary, in accordance with specified procedures, in order that the material balance can be established.

mathematical model

See *model*.

maximum normal operating pressure

The maximum pressure above atmospheric pressure at mean sea level that would develop in the *containment system* in a period of one year under the conditions of temperature and solar *radiation* corresponding to environmental conditions in the absence of venting, external cooling by an ancillary *system,* or operational *controls* during *transport*. (See SSR-6 (Rev. 1) [2].)

> ! This usage is specific to the Transport Regulations [2].

maximum potential magnitude

See *magnitude (of an earthquake)*.

mechanistic model

See *model*.

medical exposure

See *exposure categories*.

medical physicist

A *health professional* with specialist education and training in the concepts and techniques of applying physics in medicine and competent to practise independently in one or more of the subfields (specialties) of medical physics.

> ⓘ Competence of persons is normally assessed by the State by having a formal mechanism for registration, accreditation or *certification* of *medical physicists* in the various specialties (e.g. diagnostic radiology, *radiation* therapy, nuclear medicine).

> ⓘ States that have yet to develop such a mechanism would need to assess the education, training and competence of any individual proposed by the *licensee* to act as a *medical physicist* and to decide, on the basis of either international accreditation standards or standards of a State where such an accreditation system exists, whether such an individual could undertake the functions of a *medical physicist*, within the required specialty.

medical radiation facility

A medical *facility* in which *radiological procedures* are performed.

See also *facilities and activities*.

medical radiation technologist

A *health professional*, with specialist education and training in medical radiation technology, competent to perform *radiological procedures*, on delegation from the *radiological medical practitioner*, in one or more of the specialties of medical radiation technology.

> ⓘ Competence of persons is normally assessed by the State by having a formal mechanism for registration, accreditation or *certification* of *medical radiation technologists* in the various specialties (e.g. diagnostic radiology, *radiation* therapy, nuclear medicine).

> ⓘ States that have yet to develop such a mechanism would need to assess the education, training and competence of any individual proposed by the *licensee* to act as a *medical radiation technologist* and to decide, on the basis of either international standards or standards of a State where such a system exists, whether such an individual could undertake the functions of a *medical radiation technologist*, within the required specialty.

medical radiological equipment

Radiological equipment used in *medical radiation facilities* to perform *radiological procedures* that either delivers an *exposure* to an individual or directly controls or influences the extent of such exposure. The term applies to *radiation generators*, such as X ray machines or medical linear accelerators; to devices containing *sealed sources*, such as ^{60}Co teletherapy units; to devices used in a medical imaging procedure involving *ionizing radiation* to capture images, such as gamma cameras, image intensifiers or flat panel detectors; and to hybrid systems such as positron emission tomography–computed tomography scanners.

member of the public

For purposes of *protection and safety*, in a general sense, any individual in the population except when subject to *occupational exposure* or *medical exposure*. For the purpose of verifying compliance with the annual *dose limit* for *public exposure*, this is the *representative person*.

migration

The movement of radionuclides in the *environment* as a result of natural *processes*.

> ⓘ Most commonly, movement of radionuclides in association with groundwater flow.

mild environment

Environmental conditions that would at no time be significantly more severe than the conditions anticipated for *operational states*.

[mill]

See [*mine or mill processing radioactive ores*].

[milling]

See [*mining and milling*].

124

[mine or mill processing radioactive ores]

Installation for mining, [*milling*] or processing ores containing *uranium series* or *thorium series* radionuclides.

ⓘ A *mine processing radioactive ores* is any mine that yields ores containing *uranium series* or *thorium series* radionuclides, either in amounts or concentrations sufficient to warrant exploitation or, when present in conjunction with other substances being mined, in amounts or concentrations that require *radiation protection* measures to be taken as determined by the *regulatory body*.

ⓘ A *mill processing radioactive ores* is any *facility* for processing *radioactive* ores from a *mine processing radioactive ores* as here defined to produce a physical or chemical concentrate.

ⓘ This entry was restricted to those mining and processing *operations* aimed at extracting *uranium series* or *thorium series* radionuclides and those aimed at the extraction of other substances from ore where this represents a significant radiological *hazard*.

ⓘ Strictly speaking, a mill in the context of the processing of minerals is a *facility* for the processing of ore to reduce its particle size, especially by crushing or grinding. However, the term [*mill*] was used in a broader sense to denote a *facility* in which additional processing (e.g. hydrometallurgical processing) may also be carried out.

! Owing to the possibility of confusion, the use of the word [*mill*] in this broader sense, in this expression or elsewhere, is discouraged.

ⓘ This entry has been included for information only. Words are used with their usual dictionary meanings except for the term *radioactive*. See *radioactive* (2).

minimization (of waste)

The *process* of reducing the amount and *activity* of *radioactive waste* to a level as low as reasonably achievable, at all stages from the *design* of a *facility or activity* to *decommissioning*, by reducing the amount of *waste* generated and by means such as *recycling* and *reuse*, and *treatment* to reduce its *activity*, with due consideration for *secondary waste* as well as primary *waste*.

ⓘ *Minimization of waste* is not to be confused with *volume reduction*.

See *radioactive waste management*.

recycling. The process of converting *waste* materials into new products.

ⓘ *Recycling* reduces the wastage of useful materials, the use of raw materials and energy use.

ⓘ *Recycling* contributes to reducing air pollution (caused by incineration) and reducing water pollution (caused by use of landfill sites) by reducing the need for disposal of conventional waste, and also contributes to reducing emissions of greenhouse gases.

reuse. The use of an item again after it has been used before.

ⓘ *Reuse* includes conventional *reuse*, in which an item is used again to perform the same functions, and *reuse* in which an item is used again to perform a different function.

minimum detectable activity (MDA)

The *radioactivity* which, if present in a sample, produces a counting rate that will be detected (i.e. considered to be above *background*) with a certain level of confidence.

ⓘ The 'certain level of confidence' is normally set at 95%; that is, a sample containing exactly the *minimum detectable activity* will, as a result of random fluctuations, be taken to be free of *radioactivity* 5% of the time.

ⓘ The *minimum detectable activity* is sometimes referred to as the **detection limit** or **lower limit of detection**.

ⓘ The counting rate from a sample containing the *minimum detectable activity* is termed the **determination level**.

minimum significant activity (MSA)

The *radioactivity* which, if present in a sample, produces a counting rate that can be reliably distinguished from *background* with a certain level of confidence.

ⓘ A sample containing exactly the *minimum significant activity* will, as a result of random fluctuations, be taken to be free of *radioactivity* 50% of the time, whereas a true *background* sample will be taken to be free of *radioactivity* 95% of the time.

ⓘ The *minimum significant activity* is sometimes referred to as the **decision limit**. The counting rate from a sample containing the *minimum significant activity* is termed the **critical level**.

[mining and milling]

Mining in a mine that yields *radioactive* ores containing *uranium series* or *thorium series* radionuclides, either in amounts or concentrations sufficient to warrant exploitation or, when present in conjunction with other substances being mined, in amounts or concentrations that require *radiation protection* measures to be taken as determined by the *regulatory body*; and processing of *radioactive* ores from such mines to produce a chemical concentrate.

ⓘ This entry was restricted to those mining and processing *operations* aimed at extracting *uranium series* or *thorium series* radionuclides and those aimed at the extraction of other substances from ore where this represents a significant radiological *hazard*.

ⓘ Strictly speaking, milling in the context of the processing of minerals is the processing of ore to reduce its particle size, especially by crushing or grinding.

ⓘ However, in the context of this entry, the term [*milling*] was used in a broader sense to include additional processing (e.g. hydrometallurgical processing).

! Owing to the possibility of confusion, the use of the word [*milling*] in this broader sense, in this expression or elsewhere, is discouraged.

ⓘ Mining includes in situ leaching, also known as solution mining or in situ recovery, which involves recovering minerals from ores in the ground by dissolving them and pumping the resultant solution to the surface so that the minerals can be recovered.

ⓘ This entry has been included for information only. The terms mining and [*milling*] should be used with their usual dictionary meanings, qualified where necessary (e.g. by use of the term *radioactive* ores).

See also [*mine or mill processing radioactive ores*].

[mining and milling waste (MMW)]

See *waste*.

mission time

The length of time for which equipment is intended to perform its intended function in *accident conditions*.

ⓘ In the context of *probabilistic safety assessment (PSA)*, the *mission time* is a time period that a *system* or *component* is intended to operate in order to successfully perform its function.

mitigatory action

See *protective action* (1).

mixed waste

See *waste*.

model

An analytical or physical representation or quantification of a real *system* and the ways in which phenomena occur within that *system*, used to predict or assess the behaviour of the real *system* under specified (often hypothetical) conditions.

computational model. A calculational tool that implements a *mathematical model*.

conceptual model. A set of qualitative assumptions used to describe a *system* (or part thereof).

ⓘ These assumptions would normally cover, as a minimum, the geometry and dimensionality of the *system*, initial and boundary conditions, time dependence, and the nature of the relevant physical, chemical and biological *processes* and phenomena.

mathematical model. A set of mathematical equations designed to represent a *conceptual model*.

mechanistic model (biophysical model). Representation of an assumed or proven *radiation* induced biophysical *process* occurring on the molecular level, cellular level, organ level or level of the whole organism.

physical model. A physical representation, at different scale and/or using different materials, of a *structure* or *component*, the performance of which may be related to that of the real structure or component.

risk projection model. A *conceptual model* such as that for estimating the *risk* from *radiation exposure* at low *doses* and *dose rates* on the basis of epidemiological evidence concerning the *risk* from high *doses* and/or *dose rates*.

additive risk projection model. A *risk projection model* in which *exposure* is assumed to lead to an *attributable risk* that is proportional to the *dose* but independent of the natural probability of the effect.

multiplicative risk projection model. A *risk projection model* in which *exposure* is assumed to lead to an *attributable risk* that is proportional to the *dose* and to the natural probability of the effect.

seismotectonic model. A *model* that characterizes seismic sources in the region around a site of interest, including the *aleatory uncertainties* and the *epistemic uncertainties* in the seismic source characteristics.

model calibration

See *calibration*.

model validation

See *validation* (1).

model verification

See *verification* (1).

monitoring

1.   The measurement of *dose, dose rate* or *activity* for reasons relating to the *assessment* or *control* of *exposure* to *radiation* or exposure due to *radioactive substances*, and the interpretation of the results.

   ⓘ 'Measurement' is used somewhat loosely. The 'measurement' of *dose* often means the measurement of a *dose equivalent quantity* as a proxy (i.e. substitute) for a *dose quantity* that cannot be measured directly. Also, sampling may be involved as a preliminary step to measurement.

   ⓘ Measurements may actually be of *radiation* levels, airborne *activity concentrations*, levels of *contamination*, quantities of *radioactive material* or *individual doses*.

   ⓘ The results of these measurements may be used to assess radiological *hazards* or *doses* resulting or potentially resulting from *exposure*.

   ⓘ *Monitoring* may be subdivided in two different ways: according to where the measurements are made, into *individual monitoring, workplace monitoring, source monitoring* and *environmental monitoring*; and, according to the purpose of the *monitoring*, into *routine monitoring, task related monitoring* and *special monitoring*.

   area monitoring. A form of *workplace monitoring* in which an *area* is monitored by taking measurements at different points in that *area*.

   ⓘ As opposed to measurements by a static monitor.

   environmental monitoring. The measurement of external *dose rates* due to *sources* in the *environment* or of radionuclide concentrations in environmental media.

   ⓘ Contrasted with *source monitoring*.

   individual monitoring. *Monitoring* using measurements by equipment worn by individuals, or measurements of quantities of *radioactive substances* in or on, or taken into, the bodies of individuals, or measurements of quantities of *radioactive substances* excreted from the body by individuals.

   ⓘ Also called *personal monitoring*.

   ⓘ For *workers*, usually contrasted with *workplace monitoring*.

   ⓘ It includes, for example, measurements of quantities of *radioactive substances* taken into the body made using breathing zone air samplers.

   [*personal monitoring*]. Synonymous with *individual monitoring*.

   ⓘ This usage may be confusing and is discouraged in favour of *individual monitoring*.

[*personnel monitoring*]. A combination of *individual monitoring* and *workplace monitoring*.

ⓘ This usage may be confusing and is discouraged in favour of *individual monitoring* and/or *workplace monitoring*, as appropriate.

routine monitoring. *Monitoring* associated with continuing *operations* and intended: (1) to demonstrate that working conditions, including the levels of *individual dose*, remain satisfactory; and (2) to meet regulatory *requirements*.

ⓘ *Routine monitoring* can be *individual monitoring* or *workplace monitoring*.

ⓘ Contrasting terms: *task related monitoring* and *special monitoring*.

source monitoring. The measurement of *activity* in radionuclides being released to the *environment* or of external *dose rates* due to *sources* within a *facility or activity*.

ⓘ Contrasted with *environmental monitoring*.

special monitoring. *Monitoring* designed to investigate a specific situation in the workplace for which insufficient information is available to demonstrate adequate *control*, by providing detailed information to elucidate any problems and to define future *procedures*.

ⓘ *Special monitoring* would normally be undertaken at the *commissioning* stage of new *facilities*, following major modifications either to *facilities* or to *procedures*, or when *operations* are being carried out under abnormal circumstances, such as following an *accident*.

ⓘ *Special monitoring* can be *individual monitoring* or *workplace monitoring*.

ⓘ Contrasting terms: *routine monitoring* and *task related monitoring*.

task related monitoring. *Monitoring* in relation to a specific *operation*, to provide data to support immediate decisions on the management of the *operation*.

ⓘ *Task related monitoring* can be *individual monitoring* or *workplace monitoring*.

ⓘ Contrasting terms: *routine monitoring* and *special monitoring*.

workplace monitoring. *Monitoring* using measurements made in the working environment.

ⓘ Usually contrasted with *individual monitoring*.

2. Continuous or periodic measurement of radiological or other parameters or determination of the status of a *structure, system or component*.

ⓘ Sampling may be involved as a preliminary step to measurement.

ⓘ Although the concept is not fundamentally different from definition (1), this definition is more suited to the types of *monitoring* concerned primarily with *safety* (i.e. keeping *sources* under *control*) rather than with *protection* (i.e. controlling *exposure*).

ⓘ This definition is particularly relevant to *monitoring* the status of a *nuclear installation* by tracking plant variables, or *monitoring* the long term performance of a *waste disposal facility* by tracking variables such as water fluxes.

ⓘ These examples differ from definition (1) in that the routine measurements are themselves of no particular interest; the *monitoring* is only intended to detect unexpected *deviations* if they occur.

condition monitoring. Activities performed to assess the functional capability of equipment by measuring and tracking the condition of equipment.

ⓘ *Condition monitoring* is usually conducted on a non-intrusive basis.

multilateral approval

See *approval*.

multiple barriers

See *barrier*.

multiple safety functions

See *barrier*.

multiplicative risk projection model

See *model*: *risk projection model*.

national nuclear forensics library

See *nuclear forensic science*.

natural analogue

A situation in nature used as a *model* for *processes* affecting human made systems.

ⓘ The use of a *natural analogue* allows conclusions to be drawn that are relevant in making judgements about the *safety* of an existing or planned *nuclear facility*.

ⓘ In particular, mineral deposits containing radionuclides whose *migration* history over very long time periods can be analysed and the results used in modelling the potential behaviour of these or similar radionuclides in the *geosphere* over a long period of time can be used as *natural analogues*.

natural background

See *background*.

natural source

See *source* (1).

natural uranium

See *uranium*.

naturally occurring radioactive material (NORM)

Radioactive material containing no significant amounts of radionuclides other than *naturally occurring radionuclides*.

ⓘ The exact definition of 'significant amounts' would be a regulatory decision.

ⓘ Material in which the *activity concentrations* of the *naturally occurring radionuclides* have been changed by a *process* is included in *naturally occurring radioactive material (NORM)*.

ⓘ *Naturally occurring radioactive material* or *NORM* should be used in the singular unless reference is explicitly being made to various materials.

naturally occurring radionuclides

See *radionuclides of natural origin*.

near field

The excavated area of a *disposal facility* near or in contact with the *waste packages*, including filling or sealing materials, and those parts of the host medium/rock whose characteristics have been or could be altered by the *disposal facility* or its contents.

See also *far field*.

near miss

A potential significant *event* that could have occurred as the consequence of a sequence of actual occurrences but did not occur owing to the conditions prevailing at the time.

See also *event, incident* and *safety*.

near surface disposal

See *disposal* (1).

near surface disposal facility

See *disposal facility*.

need to hold

Rule by which individuals are permitted to have in their physical possession only the information assets that are necessary to conduct their work effectively.

need to know

Rule by which individuals, processes, and systems are granted access to only the information, capabilities and assets which are necessary for execution of their authorized functions.

non-fixed contamination

See *contamination* (2).

non-functional requirements

See *functional requirements*.

non-physical ageing

See *ageing*.

non-radiological consequences

Adverse psychological, societal or economic consequences of a *nuclear or radiological emergency* or of an *emergency response* affecting human life, health, property or the *environment*.

    ⓘ The definition relates to *emergency preparedness* and *response* only [22].

[non-stochastic effect]

See *health effects (of radiation)*: deterministic effect.

NORM

See *naturally occurring radioactive material*.

NORM residue

Material that remains from a *process* and comprises or is contaminated by *naturally occurring radioactive material (NORM)*.

> ⓘ A *NORM residue* may or may not be *waste*.

NORM waste

See *waste*.

normal operation

See *plant states (considered in design)*.

notification

1.   A document submitted to the *regulatory body* by a *person or organization* to notify an intention to carry out a *practice* or other use of a *source*.

> ⓘ This includes the *notification* of appropriate *competent authorities* by a *consignor* that a *shipment* will pass *through or into* their countries, as required in section V of the Transport Regulations [2].

2.   A report submitted promptly to a national or international authority providing details of an *emergency* or a possible *emergency*; for example, as required by the Convention on Early Notification of a Nuclear Accident [15].

3.   A set of actions taken upon detection of *emergency* conditions with the purpose of alerting all organizations with responsibility for *emergency response* in the event of such conditions.

notification point

A designated organization with which *arrangements* have been made to receive *notification* (meaning (3)) and to initiate promptly the predetermined actions to activate a part of the *emergency response*.

notifying State

The State that is responsible for notifying (see *notification* (2)) potentially affected States and the IAEA of an *event* of actual, potential or perceived radiological significance for other States.

> ⓘ This includes:
>
> (a) The State Party that has jurisdiction or *control* over the *facility* or *activity* (including space objects) in accordance with Article 1 of the Convention on Early Notification of a Nuclear Accident [15];
>
> (b) The State that initially detects or discovers evidence of a *transnational emergency*, for example by: detecting significant increases in atmospheric *radiation* levels of unknown origin; detecting *contamination* in transboundary *shipments*; discovering a *dangerous source* that may have originated in another State; or diagnosing clinical symptoms that may have resulted from *exposure* outside the State.

nuclear

ⓘ Strictly: relating to a nucleus; relating to or using energy released in nuclear fission or fusion. (adjective)

! The adjective *'nuclear'* is used in many phrases to modify a noun that it cannot logically modify. It must be borne in mind that the meaning of such phrases may be unclear (as opposed to *nuclear*).

! The phrases may therefore need to be explained, and their usage may be open to misunderstanding, misrepresentation or mistranslation.

! Such phrases include: *nuclear accident*; nuclear community; *nuclear emergency*; *nuclear facility*; *nuclear fuel*; *nuclear incident*; *nuclear installation*; *nuclear material*; nuclear medicine; (a) nuclear power; *nuclear safety*; and *nuclear security*.

nuclear accident

See *accident* (1).

nuisance alarm

See *alarm*.

[nuclear damage]

"(i)   [L]oss of life or personal injury;

(ii)   loss of or damage to property;

"and each of the following to the extent determined by the law of the competent court:

(iii)   economic loss arising from loss or damage referred to in sub-paragraph (i) or (ii), insofar as not included in those sub-paragraphs, if incurred by a person entitled to claim in respect of such loss or damage;

(iv)   the costs of measures of reinstatement of impaired environment, unless such impairment is insignificant, if such measures are actually taken or to be taken, and insofar as not included in sub-paragraph (ii);

(v)   loss of income deriving from an economic interest in any use or enjoyment of the environment, incurred as a result of a significant impairment of that environment, and insofar as not included in sub-paragraph (ii);

(vi)   the costs of preventive measures, and further loss or damage caused by such measures;

(vii)   any other economic loss, other than any caused by the impairment of the environment, if permitted by the general law on civil liability of the competent court,

"in the case of sub-paragraphs (i) to (v) and (vii) above, to the extent that the loss or damage arises out of or results from ionizing radiation emitted by any source of radiation inside a nuclear installation, or emitted from nuclear fuel or radioactive products or waste in, or of nuclear material coming from, originating in, or sent to, a nuclear installation, whether so arising from the radioactive properties of such matter, or from a combination of radioactive properties with toxic, explosive or other hazardous properties of such matter." (From Ref. [42].)

ⓘ In this context, 'preventive measures' are any reasonable measures taken by any person after a nuclear *incident* has occurred to prevent or minimize damage referred to in sub-paragraphs (i) to (v) or (vii), subject to any approval of the *competent authorities* required by the law of the State where the measures were taken.

nuclear emergency

See *emergency*.

nuclear facility

1.   A *facility* (including associated buildings and equipment) in which *nuclear material* is produced, processed, used, handled, stored or disposed of.

ⓘ Also called a ***nuclear fuel cycle facility***.

ⓘ A *nuclear facility* is an *authorized facility*, and this aspect of authorization is made explicit in other broadly similar definitions established in Nuclear Security Series publications, i.e. a *nuclear facility* is "A facility (including associated buildings and equipment) in which *nuclear material* is produced, processed, used, handled, stored or disposed of and for which an *authorization* or licence is required" [9] or "A facility (including associated buildings and equipment) in which *nuclear material* is produced, processed, used, handled, stored or disposed of and for which a specific licence is required" [6].

ⓘ For safeguards purposes, see the definition of *facility* in the Safeguards Glossary [14].

See also *facilities and activities, associated facility* and *nuclear installation*.

2.   [A *facility* (including associated buildings and equipment) in which *nuclear material* is produced, processed, used, handled, stored or disposed of, if damage to or interference with such *facility* could lead to the release of significant amounts of radiation or *radioactive material*.] (See Refs [4–6].)

! This usage is specific to the revised Convention on the Physical Protection of Nuclear Material and Nuclear Facilities [4–6], for the purposes of the Convention, and should otherwise be avoided.

ⓘ The 2005 Amendment to the Convention on the Physical Protection of Nuclear Material and Nuclear Facilities was adopted on 8 July 2005.

3.   ["[A] civilian facility and its associated land, buildings and equipment in which radioactive materials are produced, processed, used, handled, stored or disposed of on such a scale that consideration of safety is required".] (From Ref. [11].)

! This usage is specific to the Joint Convention on the Safety of Spent Fuel Management and on the Safety of Radioactive Waste Management [11], for the purposes of the Joint Convention, and should otherwise be avoided.

ⓘ Essentially synonymous with *authorized facility*, and hence more general than *nuclear installation*.

4.   ["(a) Any nuclear reactor, including reactors installed on vessels, vehicles, aircraft or space objects for use as an energy source in order to propel such vessels, vehicles, aircraft or space objects or for any other purpose; (b) Any plant or conveyance being used for the production, storage, processing or transport of radioactive material".]

! This usage is specific to ICSANT [12] and should otherwise be avoided.

nuclear forensic science

A discipline of forensic science involving the *examination* of nuclear or *other radioactive material*, or of other evidence that is contaminated with radionuclides, in the context of legal proceedings.

> ⓘ Also called *nuclear forensics*.

> *designated nuclear forensic laboratory.* A laboratory that has been identified by a State as being capable of accepting or analysing samples of nuclear and/or *other radioactive material* for the purpose of supporting nuclear forensic *examinations*.

> *national nuclear forensics library.* An administratively organized collection of information on nuclear and *other radioactive material* produced, used or stored within a State.

> *nuclear forensic interpretation.* The process of correlating sample characteristics with existing information on types of material, origins and methods of production of *nuclear and other radioactive material*, or with previous cases involving similar material.

nuclear fuel

Fissionable *nuclear material* in the form of fabricated elements for loading into the reactor core of a civil nuclear power plant or *research reactor*.

> *fresh fuel.* New *fuel* or *unirradiated fuel*, including *fuel* fabricated from *fissionable material* recovered by *reprocessing* previously irradiated *fuel*.

nuclear fuel cycle

All *operations* associated with the production of nuclear energy.

> ⓘ *Operations* in the *nuclear fuel cycle* associated with the production of nuclear energy include the following:
>
> (a) Mining and processing of *uranium* ores or thorium ores;
>
> (b) Enrichment of *uranium*;
>
> (c) Manufacture of *nuclear fuel*;
>
> (d) *Operation* of nuclear reactors (including *research reactors*);
>
> (e) *Reprocessing* of *spent fuel*;
>
> (f) All *waste management activities* (including *decommissioning*) relating to *operations* associated with the production of nuclear energy;
>
> (g) Any related research and development *activities*.

> *closed nuclear fuel cycle.* Mining, processing, conversion, enrichment of *uranium*, *nuclear fuel* fabrication, reactor *operation*, electrical generation or other energy products, *reprocessing* to recover *fissile material*, *storage* of reprocessed *fissile material*, *disposal* (for highly *radioactive fission products*) and final *end states* for all *waste*.

> *open nuclear fuel cycle.* Mining, processing, conversion, enrichment of *uranium*, *nuclear fuel* fabrication, reactor *operation*, electrical generation or other energy products, *storage* of *spent fuel*, *disposal* and final *end states* for all *waste*.

nuclear fuel cycle facility

See *nuclear facility*.

[nuclear incident]

See *incident*.

nuclear installation

1.   Any *nuclear facility* subject to *authorization* that is part of the *nuclear fuel cycle*, except *facilities* for the mining or processing of *uranium* ores or thorium ores and *disposal facilities* for *radioactive waste*.

> ⓘ This definition thus includes: nuclear power plants; *research reactors* (including subcritical and *critical assemblies*) and any adjoining radioisotope production *facilities*; *storage facilities* for *spent fuel*; *facilities* for the enrichment of *uranium*; *nuclear fuel* fabrication *facilities*; conversion *facilities*; *facilities* for the *reprocessing* of *spent fuel*; *facilities* for the *predisposal management* of *radioactive waste* arising from *nuclear fuel cycle facilities*; and *nuclear fuel cycle* related research and development *facilities*.

> ⓘ For safeguards purposes, see the definition of *nuclear installations* in the Safeguards Glossary [14].

2.   [For each Contracting Party, any land-based civil nuclear power plant under its jurisdiction, including such *storage*, handling and treatment *facilities* for radioactive materials as are on the same site and are directly related to the *operation* of the nuclear power plant. Such a plant ceases to be a *nuclear installation* when all nuclear fuel elements have been removed permanently from the reactor core and have been stored safely in accordance with approved procedures, and a *decommissioning* programme has been agreed to by the *regulatory body*.] (See Ref. [10].)

nuclear material

1.   Material listed in the table on the categorization of nuclear material, including the material listed in its footnotes, in Section 4 of IAEA Nuclear Security Series No. 13, Nuclear Security Recommendations on Physical Protection of Nuclear Material and Nuclear Facilities (INFCIRC/225/Revision 5) [6].

> ⓘ This table is reproduced below.

Category I/II/III nuclear material. See Table 2.

TABLE 2. CATEGORIZATION OF NUCLEAR MATERIAL (REPRODUCED FROM REF. [6])

Material	Form	Category I	Category II	Category III[c]
1. Plutonium[a]	Unirradiated[b]	2 kg or more	Less than 2 kg but more than 500 g	500 g or less but more than 15 g
2. Uranium-235 (^{235}U)	Unirradiated[b]			
	— Uranium enriched to 20% ^{235}U or more	5 kg or more	Less than 5 kg but more than 1 kg	1 kg or less but more than 15 g
	— Uranium enriched to 10% ^{235}U but less than 20%		10 kg or more	Less than 10 kg but more than 1 kg
	— Uranium enriched above natural, but less than 10% ^{235}U			10 kg or more
3. Uranium-233 (^{233}U)	Unirradiated[b]	2 kg or more	Less than 2 kg but more than 500 g	500 g or less but more than 15 g
4. Irradiated fuel (The categorization of irradiated fuel in the table is based on international transport considerations. The State may assign a different category for domestic use, storage and transport taking all relevant factors into account.)			Depleted or natural uranium, thorium or low-enriched fuel (less than 10% fissile content)[d,e]	

[a] All plutonium except that with isotopic concentration exceeding 80% in plutonium-238.

[b] Material not irradiated in a reactor or material irradiated in a reactor but with a radiation level equal to or less than 1 Gy/h (100 rad/h) at 1 m unshielded.

[c] Quantities not falling in Category III and natural uranium, depleted uranium and thorium should be protected at least in accordance with prudent management practice.

[d] Although this level of protection is recommended, it would be open to States, upon evaluation of the specific circumstances, to assign a different category of physical protection.

[e] Other fuel which by virtue of its original fissile material content is classified as Category I and II before irradiation may be reduced one category level while the radiation level from the fuel exceeds 1 Gy/h (100 rad/h) at 1 m unshielded.

2.   Any material that is either *special fissionable material* or *source material* as defined in Article XX of the IAEA Statute.

    ⓘ *Nuclear material* is necessary for the production of nuclear weapons or other nuclear explosive devices. Under comprehensive *safeguards agreements*, the IAEA verifies that all *nuclear material* subject to safeguards has been declared and placed under safeguards.

    ⓘ Certain non-nuclear materials are essential for the use or production of *nuclear material* and may also be subject to IAEA safeguards under certain agreements.

ⓘ For safeguards purposes, see the definition of *nuclear material* in the Safeguards Glossary [14].

ⓘ The Statute of the IAEA [47] uses the term *special fissionable material*, with the meaning essentially of *nuclear material* as defined here, but explicitly excluding *source material*.

special fissionable material. Plutonium-239; uranium-233; *uranium enriched in the isotopes 235 or 233*; any material containing one or more of the foregoing; and such other fissionable material as the [IAEA] Board of Governors shall from time to time determine; but not including *source material*. (See Ref. [47].)

source material. *Uranium* containing the mixture of isotopes occurring in nature; *uranium* depleted in the isotope 235; thorium; any of the foregoing in the form of metal, alloy, chemical compound, or concentrate; any other material containing one or more of the foregoing in such concentration as the [IAEA] Board of Governors shall from time to time determine; and such other material as the [IAEA] Board of Governors shall from time to time determine. (See Ref. [47].)

3.   Plutonium except that with isotopic concentration exceeding 80% in ^{238}Pu; ^{233}U; *uranium* enriched in the isotope 235 or 233; *uranium* containing the mixture of isotopes as occurring in nature other than in the form of ore or ore residue; any material containing one or more of the foregoing.

ⓘ This definition is used in the CPPNM [4, 5] and ICSANT [12].

ⓘ The 2005 Amendment to the Convention on the Physical Protection of Nuclear Material and Nuclear Facilities was adopted on 8 July 2005.

ⓘ See Refs [4–6].

ⓘ For practical purposes, all three definitions of *nuclear material* are assumed to refer to broadly the same range of material.

ⓘ The Paris Convention on Third Party Liability in the Field of Nuclear Energy [48] uses the term 'nuclear substances', which means *nuclear fuel* (other than *natural uranium* and *depleted uranium*) and *radioactive products* or *radioactive waste*.

nuclear material control

See *control (of nuclear material)*.

nuclear or radiological emergency

See *emergency*.

(nuclear) safety

The achievement of proper *operating conditions*, prevention of *accidents* and mitigation of *accident* consequences, resulting in *protection* of *workers*, the public and the *environment* from undue *radiation risks*.

ⓘ Often abbreviated to *safety* in *IAEA publications* on *nuclear safety*. *Safety* means *nuclear safety* unless otherwise stated, in particular when other types of *safety* (e.g. fire *safety*, conventional industrial *safety*) are also being discussed.

ⓘ *Safety* encompasses the *safety* of *nuclear installations*, radiation *safety*, the *safety* of *radioactive waste management* and *safety* in the *transport* of *radioactive material*.

ⓘ There is not an exact distinction between the general terms *safety* and *security*. In general, *security* is concerned with intentional actions by people that could cause or threaten harm to other people; *safety* is concerned with the broader issue of harmful consequences to people (and to the *environment*) arising from *exposure* to *radiation*, whatever the cause.

ⓘ The interaction between arrangements for *security* and arrangements for *safety* depends on the context. *Areas* in which arrangements for *safety* and arrangements for *security* interact include, for example: the regulatory infrastructure; engineering provisions in the *design* and *construction* of *nuclear installations* and other *facilities*; *controls* on access to *nuclear installations* and other *facilities*; the categorization of *radioactive sources*; *source design*; the *security* of the management of *radioactive sources* and *radioactive material*; the recovery of *sources* that are not under *regulatory control*; *emergency response* plans; and *radioactive waste management*.

See *protection and safety* for a discussion of the relationship between *nuclear safety* and *radiation protection*.

nuclear security

1.   The prevention and detection of, and *response* to, criminal or intentional unauthorized acts involving or directed at *nuclear material, other radioactive material, associated facilities* or *associated activities*.

ⓘ Often abbreviated to *security* in *IAEA publications* on *nuclear security*.

ⓘ 'Security' in a general sense encompasses related issues of global security — the sustainability of human life — in terms of energy security, environmental security, *food* security and water security, as well as *nuclear security* — to all of which the use of nuclear energy is related.

ⓘ *Security* of *nuclear material* for reasons relating to non-proliferation of nuclear weapons is outside the scope of the IAEA *safety standards* and of the IAEA Nuclear Security Series.

ⓘ There is not an exact distinction between the general terms *safety* and *security*. In general, *security* is concerned with intentional actions by people that could cause or threaten harm to persons, property, society or the environment; *safety* is concerned with the broader issue of harmful consequences to people (and to the *environment*) arising from *exposure* to *radiation*, whatever the cause.

ⓘ The interaction between arrangements for *security* and arrangements for *safety* depends on the context. *Areas* in which arrangements for *safety* and arrangements for *security* interact include, for example: the regulatory infrastructure; engineering provisions in the *design* and *construction* of *nuclear installations* and other *facilities*; *controls* on access to *nuclear installations* and other *facilities*; the categorization of *radioactive sources*; *source design*; the *security* of the *radioactive sources* and *radioactive material*; the recovery of *sources* that are not under *regulatory control*; *emergency response* plans; and *radioactive waste management*.

2.   [The prevention and detection of, and *response* to, theft, sabotage, unauthorized access, illegal transfer or other malicious acts involving *nuclear material, other radioactive material* or their *associated facilities*.]

ⓘ This definition is taken from the second Nuclear Security Plan (IAEA GOV/2005/50).

nuclear security culture

The assembly of characteristics, attitudes and behaviours of individuals, organizations and institutions which serves as a means to support, enhance and sustain *nuclear security*.

See also *safety culture*.

nuclear security detection architecture

The integrated set of nuclear security systems and measures as defined in IAEA Nuclear Security Series No. 15, Nuclear Security Recommendations on Nuclear and Other Radioactive Material out of Regulatory Control based on an appropriate legal and regulatory framework needed to implement the national strategy for the detection of nuclear and *other radioactive material out of regulatory control*.

nuclear security event

An *event* that has potential or actual implications for *nuclear security* that must be addressed.

    ⓘ In the context of *authorized facilities* and *authorized activities*, the following definition has also been established in some publication of the Nuclear Security Series: "An event that is assessed as having implications for *nuclear security*" [6, 7].

    ⓘ Such *events* include criminal or intentional unauthorized acts involving or directed at *nuclear material, other radioactive material, associated facilities* or *associated activities*.

    ⓘ A *nuclear security event*, for example, *sabotage* of a *nuclear facility* or detonation of a *radiological dispersal device*, may give rise to a *nuclear or radiological emergency*.

nuclear security measures

Measures intended to prevent a *nuclear security threat* from completing criminal or intentional unauthorized acts involving or directed at *nuclear material, other radioactive material, associated facilities*, or *associated activities* or to detect or respond to *nuclear security events*.

    ⓘ In some Nuclear Security Series publications, this term has also been defined in a broadly similar way as: "Measures intended to prevent a *threat* from completing a *malicious act* or to *detect* or respond to *nuclear security events*" [7].

nuclear security regime

A regime comprising:

- The legislative and regulatory framework and administrative systems and measures governing the nuclear security of *nuclear material, other radioactive material, associated facilities*, and *associated activities*,

- The institutions and organizations within the State responsible for ensuring the implementation of the legislative and regulatory framework and administrative systems of nuclear security;

- Nuclear *security systems* and *nuclear security measures* for the prevention of, detection of, and *response* to, *nuclear security events*.

nuclear security system

An integrated set of *nuclear security measures*.

O

observed cause

See *cause*.

occupancy factor

A typical fraction of the time for which a location is occupied by an individual or group.

occupational exposure

See *exposure categories*.

off-site (area)

Outside the *site area*.

on-site (area)

Within the *site area*.

open nuclear fuel cycle

See *nuclear fuel cycle*.

operating conditions

See *plant states (considered in design)*: *operational states*.

operating lifetime, operating life

See *life, lifetime*.

operating organization

1.   Any organization or person applying for *authorization* or authorized to operate an *authorized facility* or to conduct an *authorized activity* and responsible for its *safety*.

> ! Note that such an organization may be the *operating organization* before *operation* starts.

> ⓘ This includes, inter alia, private individuals, governmental bodies, *consignors* or *carriers*, *licensees*, hospitals and self-employed persons.

> ⓘ *Operating organization* includes either those who are directly in control of a *facility* or an *activity* during use of a *source* (such as radiographers or carriers) or, in the case of a *source* not under *control* (such as a lost or illicitly removed source or a re-entering satellite), those who were responsible for the *source* before *control* over it was lost.

> ⓘ In practice, for an *authorized facility*, the *operating organization* is normally also the *registrant* or *licensee*. However, the separate terms are retained to refer to the two different capacities.

See also *operator*.

2.   The organization (and its contractors) which undertakes the *siting, design, construction, commissioning* and/or *operation* of a *nuclear facility*.

> ! This usage is particular to documentation relating to the *safety* of *radioactive waste management*, with the corresponding understanding of *siting* as a multistage *process*.

> ! This difference is partly a reflection of the particularly crucial role of *siting* in the *safety* of *repositories*.

operating period

See *life, lifetime: operating lifetime, operating life* (1).

operating personnel

Individual *workers* engaged in the *operation* of an *authorized facility* or the conduct of an *authorized activity*.

> ! This may be shortened to *operator(s)*, provided that there is no danger of confusion with *operator* in the sense of *operating organization*.

operation

All *activities* performed to achieve the purpose for which an *authorized facility* was constructed.

> ⓘ For a nuclear power plant, this includes *maintenance*, refuelling, *in-service inspection* and other related *activities*.

> ⓘ The terms *siting, design, construction, commissioning, operation* and *decommissioning* are normally used to delineate the six major stages of the *lifetime* of an *authorized facility* and of the associated *licensing process*. In the special case of *disposal facilities* for *radioactive waste, decommissioning* is replaced in this sequence by *closure*.

See also *abnormal operation* and *normal operation*.

operational bypass

See *bypass* (1).

operational control area

See *area*.

operational criteria

Values of measurable quantities or observable conditions (i.e. observables) to be used in the *response* to a *nuclear or radiological emergency* in order to determine the need for appropriate *protective actions* and *other response actions*.

> ⓘ *Operational criteria* used in an *emergency* include *operational intervention levels (OILs), emergency action levels (EALs)*, specific observable conditions (i.e. observables) and other indicators of conditions on the site.

> ⓘ *Operational criteria* are sometimes referred to as triggers.

operational intervention level (OIL)

See *level*.

operational limits and conditions

See *limit*.

operational period

See *life, lifetime*: *operating lifetime, operating life* (1).

operational quantities

Quantities used in practical applications for *monitoring* and investigations that involve *external exposure*.

- ⓘ *Operational quantities* are defined for the purpose of measurement and *assessment* of *doses* in the human body.

- ⓘ In internal dosimetry, no operational *dose quantities* have been defined that directly provide an *assessment* of *equivalent dose* or *effective dose*.

- ⓘ Different methods are applied to assess the *equivalent dose* or *effective dose* from *exposure* due to radionuclides in the human body.

- ⓘ These methods are mostly based on various activity measurements and the application of biokinetic *models (computational models)*.

- ⓘ It is possible to use the measurable properties of radiation fields and of radionuclides associated with *external exposure* or with *intake* of radionuclides to estimate *protection quantities* and to demonstrate compliance with *requirements* involving *protection quantities*. These measurable quantities are called *operational quantities*.

operational states

See *plant states (considered in design)*.

operations area

See *area*.

operations boundary

See *area*: *operations area*.

operator

Any *person or organization* applying for *authorization* or authorized and/or responsible for *safety* when undertaking *activities* or in relation to any *nuclear facilities* or *sources* of *ionizing radiation*.

- ⓘ In some publications in the Nuclear Security Series, this term has been also defined as "Any person, organization, or government entity licensed or authorized to undertake the operation of an *associated facility* or to perform an *associated activity*" [6].

- ⓘ *Operator* includes, inter alia, private individuals, governmental bodies, *consignors* or *carriers*, *licensees*, hospitals, self-employed persons.

! *Operator* is sometimes used to refer to *operating personnel* (e.g. control room operators). If used in this way, particular care should be taken to ensure that there is no possibility of confusion.

ⓘ *Operator* includes either those who are directly in *control* of a *facility* or an *activity* during use or *transport* of a *source* (such as radiographers or *carriers*) or, in the case of a *source* not under *control* (such as a lost or illicitly removed *source* or a re-entering satellite), those who were responsible for the *source* before *control* over it was lost.

ⓘ Synonymous with *operating organization*.

optimization (of protection and safety)

1.   The *process* of determining what level of *protection and safety* would result in the magnitude of *individual doses*, the number of individuals (*workers* and *members of the public*) subject to *exposure* and the likelihood of *exposure* being *as low as reasonably achievable*, economic and social factors being taken into account (*ALARA*).

2.   The management of the radiation *dose* to the *patient* commensurate with the medical purpose.

ⓘ For *medical exposures* of *patients*.

ⓘ '*Optimization of protection and safety* has been implemented' means that *optimization of protection and safety* has been applied and the results of that *process* have been implemented.

! This is not the same as optimization of the *process* or *practice* concerned. An explicit term such as *optimization of protection and safety* should be used.

! The acronym *ALARA* should not be used to mean *optimization of protection and safety*.

organ dose

See *dose quantities*.

orphan source

See *source* (2).

other nuclear or radiological emergency

See *emergency class*.

other radioactive material

Any radioactive material that is not nuclear material.

other response actions

See *emergency response*: *emergency response action*.

out of regulatory control

See *regulatory control*.

overpack

1.   See *radioactive waste management* (1).

2.   An enclosure used by a single *consignor* to contain one or more *packages,* and to form one unit for convenience of handling and stowage during *transport.* (See SSR-6 (Rev. 1) [2].)

package

The complete product of the packing operation, consisting of the *packaging* and its contents prepared for *transport*. The types of *package* covered by [the Transport] Regulations [2] that are subject to the *activity limits* and material restrictions of Section IV [of the Transport Regulations [2]] and meet the corresponding *requirements* are:

(a)   Excepted package;

(b)   Industrial package Type 1 (Type IP-1);

(c)   Industrial package Type 2 (Type IP-2);

(d)   Industrial package Type 3 (Type IP-3);

(e)   Type A package;

(f)   Type B(U) package;

(g)   Type B(M) package;

(h)   Type C package.

Packages containing *fissile material* or *uranium* hexafluoride are subject to additional *requirements*. (See SSR-6 (Rev. 1) [2].)

ⓘ The detailed specifications and *requirements* for these *package* types are specified in SSR-6 (Rev. 1) [2].

package, waste

See *waste package*.

packaging

1.   One or more receptacles and any other *components* or materials necessary for the receptacles to perform *containment* and other *safety functions*. (See SSR-6 (Rev. 1) [2].)

2.   See *radioactive waste management* (1).

palaeoseismicity

The evidence of a prehistoric or historical earthquake manifested as displacement on a fault or secondary effects such as ground deformation (i.e. liquefaction, tsunami, landslides).

particle fluence

See *fluence*.

passenger aircraft

See *aircraft*.

passive component

A *component* whose functioning does not depend on an external input such as actuation, mechanical movement or supply of power.

    ⓘ A *passive component* has no moving part, and, for example, only experiences a change in pressure, in temperature or in fluid flow in performing its functions. In addition, certain *components* that function with very high *reliability* based on irreversible action or change may be assigned to this category.

    ⓘ Examples of *passive components* are heat exchangers, pipes, vessels, electrical cables and *structures*. It is emphasized that this definition is necessarily general in nature, as is the corresponding definition of *active component*.

    ⓘ Certain *components*, such as rupture discs, check valves, *safety* valves, injectors and some solid state electronic devices, have characteristics which require special consideration before designation as an *active component* or a *passive component*.

    ⓘ Any *component* that is not a *passive component* is an *active component*.

See also *component, core components* and *structures, systems and components*.

pathway

See *exposure pathway*.

patient

An individual who is a recipient of services of *health professionals* and/or their agents that are directed at (a) promotion of health; (b) prevention of illness and injury; (c) monitoring of health; (d) maintaining health; and (e) medical treatment of diseases, disorders and injuries in order to achieve a cure or, failing that, optimum comfort and function. Some asymptomatic individuals are included.

    ⓘ For the purpose of the *requirements* on *medical exposure* in the IAEA *safety standards*, the term '*patient*' refers only to those individuals undergoing radiological procedures.

peak ground acceleration

The maximum absolute value of ground acceleration displayed on an *accelerogram*; the greatest ground acceleration produced by an earthquake at a site.

peer review

An *examination* or review of commercial, professional or academic efficiency, competence, etc., by others in the same occupation.

    ⓘ *Peer review* is also: the evaluation, by experts in the relevant field, of a scientific research project for which a grant is sought; the process by which a learned journal passes a paper received for publication to outside experts for their comments on its suitability and worth; refereeing.

performance assessment

See *assessment* (1).

performance indicator

See *indicator*.

performance standard

Description of the performance required of a *structure, system or component* or other item of equipment, a person or a *procedure* with the aim of ensuring a high level of *safety*.

performance testing

Testing of the *physical protection measures* and the *physical protection system* to determine whether or not they are implemented as designed; adequate for the proposed natural, industrial and *threat* environments; and in compliance with established performance requirements.

ⓘ This definition is for use in the context of *physical protection* of *nuclear material* and *nuclear facilities* and other *radioactive material* and *associated facilities and activities*.

See also *performance assessment*.

periodic maintenance

See *maintenance*.

periodic safety review

A systematic reassessment of the *safety* of an existing *facility (or activity)* carried out at regular intervals to deal with the cumulative effects of *ageing*, modifications, operating experience, technical developments and *siting* aspects, and aimed at ensuring a high level of *safety* throughout the *service life* of the *facility (or activity)*.

permanent relocation

See *relocation*.

permanent shutdown

See *shutdown*.

person or organization

Any organization, corporation, partnership, firm, association, trust, estate, public or private institution, group, political or administrative entity, or other persons designated in accordance with national legislation who or which has responsibility and authority for any action having implications for *protection and safety*.

ⓘ Supersedes the term *legal person*, which is contrasted in legal texts with 'natural person', meaning an individual.

personal dose equivalent, $H_p(d)$

See *dose equivalent quantities*.

[personal monitoring]

See *monitoring* (1).

[personnel monitoring]

See *monitoring* (1).

phreatic eruption

See *volcanic eruption*.

phreatomagmatic eruption

See *volcanic eruption*.

physical ageing

See *ageing*.

physical barrier

A fence, wall or similar impediment which provides *access delay* and complements access control.

> § See also '*barrier*'.

physical diversity

See *diversity*.

physical half-life

See *half-life* (2): *radioactive half-life*.

physical inventory

See *inventory*.

physical model

See *model*.

physical protection

Measures for the protection of *nuclear material* or authorized *facilities*, designed to prevent unauthorized access or removal of *fissile material* or sabotage with regard to safeguards, as, for example, in the Convention on the Physical Protection of Nuclear Material.

> ⓘ This definition is used in the CPPNM [4, 5].

> ⓘ The 2005 Amendment to the Convention on the Physical Protection of Nuclear Material and Nuclear Facilities was adopted on 8 July 2005.

> ⓘ The preferred approach, in cases where the term *physical protection* is still used (i.e. primarily in guidance supporting Ref. [6]), is to use it without explicit definition, making its meaning clear from the context and the measures described.

ⓘ A footnote in Ref. [6] effectively defines physical protection as the nuclear security of nuclear material and nuclear facilities. Hence, when the context is clearly *nuclear material* and *nuclear facilities*, physical protection and *nuclear security* may be considered synonymous. However, the term physical protection is sometimes understood to exclude 'non-physical' security measures, such as *computer security* or *nuclear material accounting and control*, so the preference is to avoid explicit definitions.

physical protection measures. The personnel, procedures, and equipment that constitute a *physical protection system.*

physical protection regime. A State's regime including:

- the legislative and regulatory framework governing the physical protection of *nuclear material* and *nuclear facilities*;

- the institutions and organizations within the State responsible for ensuring implementation of the legislative and regulatory framework;

- facility and transport *physical protection systems.*

ⓘ Because this is an established term, it may be used where essential for consistency with Ref. [6], with the understanding that a *physical protection regime* is that part of a State's *nuclear security* regime intended to counter unauthorized removal and sabotage of nuclear material and sabotage of nuclear facilities. However, guidance should not normally refer to a 'regime' covering a part of *nuclear security* (e.g. a transport security regime): the convention is that a State has one *national nuclear security* regime, which includes elements relating to particular areas of *nuclear security*.

physical protection system. An integrated set of *physical protection measures* intended to prevent the completion of a *malicious act.*

physical protection measures

See *physical protection.*

physical protection regime

See *physical protection.*

physical protection system

See *physical protection.*

physical separation

Separation by geometry (distance, orientation, etc.), by appropriate *barriers*, or by a combination thereof.

physisorption

See *sorption.*

planned exposure situation

See *exposure situations.*

planned maintenance

See *maintenance*.

planning target volume

A geometrical concept used in *radiation* therapy for planning medical treatment with consideration of the net effect of movements of the *patient* and of the tissues to be irradiated, variations in size and shape of the tissues, and variations in beam geometry such as beam size and beam direction.

plant equipment (for a nuclear power plant)

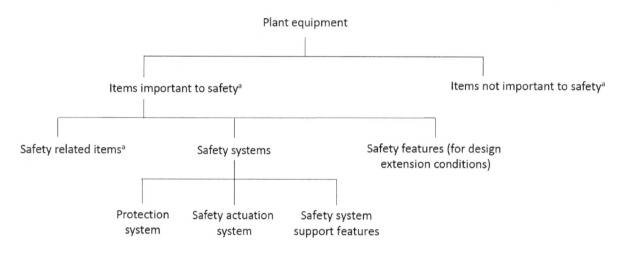

a In this context, an 'item' is a structure, system or component.

FIG. 3. Plant equipment for a nuclear power plant.

item important to safety. An item that is part of a *safety group* and/or whose malfunction or *failure* could lead to *radiation exposure* of the *site personnel* or *members of the public*.

ⓘ *Items important to safety* include:

— Those *structures, systems and components* whose malfunction or *failure* could lead to undue *radiation exposure* of *site personnel* or *members of the public*;

— Those *structures, systems and components* that prevent *anticipated operational occurrences* from leading to *accident conditions*;

— *Safety features for design extension conditions*;

— Those features that are provided to mitigate the consequences of malfunction or *failure* of *structures, systems and components*.

protection system. *System* that monitors the *operation* of a reactor and which, on sensing an abnormal condition, automatically initiates actions to prevent an unsafe or potentially unsafe condition.

! This use of the term *protection* refers to *protection* of the plant (*protection* (2)).

ⓘ The *system* in this case encompasses all electrical and mechanical devices and circuitry, from sensors to *actuation device* input terminals.

safety actuation system. The collection of equipment required to accomplish the necessary *safety actions* when initiated by the *protection system*.

safety feature for design extension conditions. Item that is designed to perform a *safety function* for or that has a *safety function* for *design extension conditions*.

ⓘ The concept of *safety features for design extension conditions* also applies for *research reactors* and *nuclear fuel cycle facilities*.

safety related item. An *item important to safety* that is not part of a *safety system or a safety feature for design extension conditions*.

safety related system. A *system* important to *safety* that is not part of a *safety system or a safety feature for design extension conditions*.

ⓘ The reactor coolant system, for example, is an *item important to safety* that is neither a *safety system* nor a *safety feature for design extension conditions*.

safety system. A *system* important to *safety*, provided to ensure the safe *shutdown* of the reactor or the *residual heat* removal from the reactor core, or to limit the consequences of *anticipated operational occurrences* and *design basis accidents*.

ⓘ *Safety systems* consist of the *protection system*, the *safety actuation systems* and the *safety system support features*.

ⓘ *Components* of *safety systems* may be provided solely to perform *safety functions*, or may perform *safety functions* in some plant *operational states* and non-safety functions in other *operational states*.

safety system settings. Settings for *levels* at which *safety systems* are automatically actuated in the event of *anticipated operational occurrences* or *design basis accidents*, to prevent *safety limits* from being exceeded.

safety system support features. The collection of equipment that provides services such as cooling, lubrication and energy supply required by the *protection system* and the *safety actuation systems*.

! After an *initiating event*, some required *safety system support features* may be initiated by the *protection system* and others may be initiated by the *safety actuation systems* they serve; other required *safety system support features* may not need to be initiated if they are in *operation* at the time of the *initiating event*.

plant states (considered in design)

! The entries that follow (terms and definitions) relate to consideration at the *design* stage (i.e. by means of hypothetical scenarios).

! Care needs to be taken to select, use and relate defined terms and other words in such a way that clear distinctions are drawn and may be inferred between, for example: *events* and situations (see the entry for *event*); *accidents* and other *incidents*; what is actual (i.e. what is), possible (i.e. what might be) or potential (i.e. what could become), and what is hypothetical (i.e. what is postulated or assumed); and what is observed or determined objectively, and what is decided or declared subjectively.

! 'Conditions', for example, is used in terms in the sense of rules set in *design* (as in *operational limits and conditions*) and also circumstances of *operation* (as in plant conditions); and in terms used in both *design* and *operation* (e.g. in accident conditions, service conditions).

! Drafters and reviewers thus need to bear in mind whether text concerns *design* or *operation*, or both. The potential, the postulated or the assumed in *design* needs to be distinguished from the observed or the determined

153

in *operation*; and the decided on or declared (such as an *emergency*), in both *design* and *operation*, needs to be distinguished from the former (i.e. the potential, the postulated, the assumed, the observed and the determined).

ⓘ The concept of *facility states* as it is used in the *safety standards* for *research reactors* and for *nuclear fuel cycle facilities* is broadly equivalent to the concept of *plant states* for nuclear power plants. Unless otherwise indicated, the definitions of terms grouped under '*plant states*' apply for nuclear power plants, *research reactors* and *nuclear fuel cycle facilities*.

See also *event, model, probabilistic safety assessment, uncertainty*.

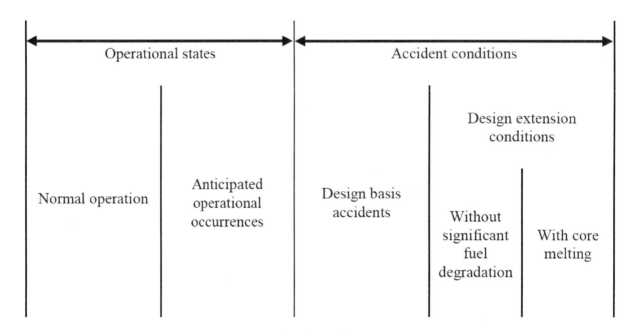

FIG. 4. Plant states considered in design for a nuclear power plant.

accident conditions. *Deviations* from *normal operation* that are less frequent and more severe than *anticipated operational occurrences*.

ⓘ *Accident conditions* are considered in the design and comprise *design basis accidents* and *design extension conditions*.

ⓘ Examples of such *deviations* include a major *fuel failure* or a loss of coolant accident (LOCA).

See also *accident* and *event*.

anticipated operational occurrence. A deviation of an operational *process* from *normal operation* that is expected to occur at least once during the *operating lifetime* of a *facility* but which, in view of appropriate *design* provisions, does not cause any significant damage to *items important to safety* or lead to *accident conditions*.

ⓘ Examples of *anticipated operational occurrences* are loss of normal electrical power and faults such as a turbine trip, malfunction of individual items of a normally running plant, *failure* to function of individual items of *control* equipment, and loss of power to the main coolant pump.

ⓘ Some States and organizations use the term **abnormal operation** (for contrast with *normal operation*) for this concept.

beyond design basis accident. Postulated *accident* with *accident conditions* more severe than those of a *design basis accident*.

154

controlled state. *Plant state*, following an *anticipated operational occurrence* or *accident conditions*, in which fulfilment of the *fundamental safety functions* can be ensured and which can be maintained for a time sufficient to implement provisions to reach a *safe state*.

design basis accident. A postulated accident leading to *accident conditions* for which a *facility* is designed in accordance with established *design* criteria and conservative methodology, and for which *releases* of *radioactive material* are kept within *acceptable limits*.

design extension conditions. Postulated *accident conditions* that are not considered for *design basis accidents*, but that are considered in the *design process* of the *facility* in accordance with best estimate methodology, and for which *releases* of *radioactive material* are kept within *acceptable limits*.

   (i) For nuclear power plants and *research reactors*, *design extension conditions* comprise conditions in *events* without significant fuel degradation and conditions in *events* with melting of the reactor core.

normal operation. *Operation* within specified *operational limits and conditions*.

   (i) For a nuclear power plant, this includes startup, power *operation*, shutting down, *shutdown*, *maintenance*, testing and refuelling.

operational states. States defined under *normal operation* and *anticipated operational occurrences*.

   (i) Some States and organizations use the term ***operating conditions*** (in contrast to *accident conditions*) for this concept.

safe state. *Plant state*, following an *anticipated operational occurrence* or *accident conditions*, in which the reactor is subcritical and the *fundamental safety functions* can be ensured and maintained stable for a long time.

plinian eruption

See volcanic *eruption*.

Pliocene

An interval of geological time extending from 5.3 to 2.6 million years ago.

point of [exit or] entry

designated point of exit or entry: An officially designated place on the land border between two States, seaport, international airport or other point where travellers, means of transport, and/or goods are inspected. Often, customs and immigration facilities are provided at these points of exit and entry.

undesignated point of exit or entry: Any air, land and water crossing point between two States that is not officially designated for travellers and/or goods by the State, such as green borders, seashores and local airports.

   (i) Sometimes referred to as border crossing points.

   ! These terms imply entry to and exit from a State and should not be confused with checkpoints or access control points that might operate at points of entry to and exit from a site, a facility or a designated area.

poison

A substance used to reduce *reactivity* (typically in a reactor core), by virtue of its high neutron *absorption* cross-section.

> [**burnable poison**]. A *poison* that becomes less effective as a result of absorbing neutrons.

> ! The term *burnable absorber* is preferred.

postulated initiating event (PIE)

See *initiating event*.

potential alpha energy

The total alpha energy ultimately emitted during the decay of decay products of ^{222}Rn or ^{220}Rn through the decay chain.

> ! Note that the definition of *radon* decay products includes the decay chain up to but not including ^{210}Pb.

> **potential alpha energy exposure.** The time integral of the *potential alpha energy* concentration in air over the time period for which an individual is exposed to radiation from decay products of ^{222}Rn or ^{220}Rn.

> ! This is not a type of *potential exposure*.

> ⓘ Used in measuring *exposure* due to decay products of ^{222}Rn or ^{220}Rn, in particular for *occupational exposure*.

> ⓘ Unit: J·h/m^3.

potential exposure

Prospectively considered *exposure* that is not expected to be delivered with certainty but that may result from an *anticipated operational occurrence* or *accident* at a *source* or owing to an *event* or sequence of *events* of a probabilistic nature, including equipment *failures* and operating errors.

> ! *Potential exposure* is not an *exposure* and is not a type of *exposure*.

> ⓘ *Potential exposure* is considered within *planned exposure situations*.

> ⓘ *Potential exposure* includes prospectively considered (i.e. hypothetical or postulated) *exposures* due to a *source* in an *event* or sequence of *events* of a probabilistic nature, including *exposures* resulting from an *accident*, equipment *failures*, operating errors, natural events or phenomena (such as hurricanes, earthquakes and floods) and inadvertent *human intrusion* (such as a *human intrusion* into a near surface *disposal facility* after *institutional control* is removed).

> ⓘ In the case of a geological *disposal facility*, *assessment* of the long term action of *processes* and *events* that are uncertain leads to projections of long term *potential exposure*.

practical elimination

ⓘ A definition of *practical elimination* is in preparation. *Practical elimination* can be achieved by, for example, the implementation of *safety* provisions, in the form of *design* and operational features, either to ensure that plant *event* sequences that could lead to an *early radioactive release* or a *large radioactive release* are physically impossible or to demonstrate, with a high level of confidence, that such plant *event* sequences are extremely unlikely to arise.

ⓘ The concept of *practical elimination* is applied in relation to plant *event* sequences, the consequences of which cannot be mitigated by reasonably practicable means.

ⓘ Practical elimination is part of a general approach to *design safety* and is an enhancement of the application of the concept of *defence in depth*.

! The phrase 'practically eliminated' is misleading as it actually concerns the possible exclusion of *event* sequences from hypothetical *scenarios* rather than practicalities of *safety*. The phrase can also all too readily be misinterpreted, misrepresented or mistranslated as referring to the 'elimination' of '*accidents*' by practical measures (or else 'practically' in the sense of 'almost'). Clear drafting in natural language would be preferable.

practice

Any human activity that introduces additional *sources* of *exposure* or additional *exposure pathways*, or that modifies the network of *exposure pathways* from existing *sources*, so as to increase the *exposure* or the likelihood of *exposure* of people or the number of people exposed.

! *Radioactive waste* is generated as a result of *practices* that involve some beneficial effect, such as the generation of electricity by nuclear means or the diagnostic application of radioisotopes. The management of this *waste* is therefore only one part of the overall *practice*.

ⓘ The term *facilities and activities* is intended to provide an alternative to the terminology of *sources* and *practices* (or *interventions*) to refer to general categories of situations.

ⓘ Terms such as 'authorized *practice*', 'controlled *practice*' and 'regulated *practice*' are used to distinguish those *practices* that are subject to *regulatory control* from other *activities* that meet the definition of a *practice* but do not need or are not amenable to *control*.

precautionary action zone (PAZ)

See *emergency planning zones*.

precautionary urgent protective action

See *protective action*: *urgent protective action*.

predeveloped block

Predeveloped functional block usable in a *hardware description language*.

ⓘ *Predeveloped blocks* include, for example, libraries, macros or intellectual property cores. A *predeveloped block* may need significant work before incorporation in a hardware programmed device.

predeveloped item

Item that already exists, is available as a commercial or proprietary product, and is being considered for use in an instrumentation and control system.

> ⓘ Predeveloped items include hardware devices, predeveloped software, commercial off the shelf devices, digital devices composed of both hardware and software, or hardware devices configured with hardware definition language or *predeveloped blocks*.

predictive maintenance

See *maintenance*.

predisposal management (of waste)

See *radioactive waste management* (1).

preferred power supply

The power supply from the transmission system to the safety classified electrical power system.

preparedness stage

See *emergency preparedness*.

[prescribed limit]

See *limit*.

pretreatment (of waste)

See *radioactive waste management* (1).

preventive maintenance

See *maintenance*.

primary limit

See *limit*.

prime mover

A *component* that converts energy into action when commanded by an *actuation device*.

> ⓘ Such as a motor, solenoid operator or pneumatic operator.

probabilistic analysis

ⓘ *Probabilistic analysis* is often taken to be synonymous with **stochastic analysis**. Strictly, however, 'stochastic' conveys directly the idea of randomness (or at least apparent randomness), whereas 'probabilistic' is directly related to probabilities, and hence only indirectly concerned with randomness.

ⓘ A natural *event* or *process* might more correctly be described as 'stochastic' (as in *stochastic effect*), whereas 'probabilistic' would be more appropriate for describing a mathematical *analysis* of *stochastic events* or *processes* and their consequences (such an *analysis* would, strictly, only be 'stochastic' if the analytical method itself included an element of randomness, e.g. Monte Carlo *analysis*).

probabilistic safety assessment (PSA)

A comprehensive, structured approach to identifying *failure scenarios*, constituting a conceptual and mathematical tool for deriving numerical estimates of *risk*.

ⓘ Three levels of *probabilistic safety assessment* are generally recognized:

— Level 1 comprises the *assessment* of *failures* leading to determination of the frequency of fuel damage.

— Level 2 includes the *assessment* of *containment* response, leading, together with Level 1 results, to the determination of frequencies of *failure* of the *containment* and *release* to the *environment* of a given percentage of the reactor core's inventory of radionuclides.

— Level 3 includes the *assessment* of *off-site* consequences, leading, together with the results of Level 2 *analysis*, to estimates of public *risks*.

(See, for example, Ref. [49].)

'living' probabilistic safety assessment. A *probabilistic safety assessment* that is updated as necessary to reflect the current *design* and operational features, and is documented in such a way that each aspect of the *PSA model* can be directly related to existing plant information and plant documentation, or to the analysts' assumptions in the absence of such information.

procedure

A series of specified actions conducted in a certain order or manner.

ⓘ The set of actions to be taken to conduct an *activity* or to perform a *process* is typically specified in a set of instructions.

process

1.   A course of action or proceeding, especially a series of progressive stages in the manufacture of a product or some other *operation*.

2.   A set of interrelated or interacting *activities* that transforms inputs into outputs.

ⓘ A product is the result or output of a *process*.

processing (of waste)

See *radioactive waste management* (1).

projected dose

See *dose concepts*.

protected area

See *area*.

protection

1.   (against *radiation*):

       radiation protection (also **radiological protection**). The *protection* of people from harmful effects of *exposure* to *ionizing radiation*, and the means for achieving this.

       See also *protection and safety*.

       ⓘ The International Commission on Radiological Protection and others use the term *radiological protection*, which is synonymous.

       ⓘ The accepted understanding of the term *radiation protection* is restricted to *protection* of people. Suggestions to extend the definition to include the *protection* of non-human species or the *protection of the environment* are controversial.

2.   (of a nuclear reactor). See *plant equipment (for a nuclear power plant): protection system*.

3.   (of *nuclear material*). See physical protection.

protection and safety

The *protection* of people against *exposure* to *ionizing radiation* or *exposure* due to *radioactive material* and the *safety* of *sources*, including the means for achieving this, and the means for preventing *accidents* and for mitigating the consequences of *accidents* if they do occur.

       ⓘ *Safety* is primarily concerned with maintaining *control* over *sources*, whereas *(radiation) protection* is primarily concerned with controlling *exposure* to *radiation* and its effects.

       ⓘ Clearly the two are closely connected: *radiation protection* (or *radiological protection*) is very much simpler if the *source* in question is under *control*, so *safety* necessarily contributes towards *protection*.

       ⓘ *Sources* come in many different types, and hence *safety* may be termed the *safety* of *nuclear installations*, *radiation safety*, the *safety* of *radioactive waste management* or *safety* in the *transport* of *radioactive material*, but *protection* (in this sense) is primarily concerned with protecting people against *exposure*, whatever the *source*, and so is always *radiation protection*.

       ⓘ For the purposes of the IAEA *safety standards*, *protection and safety* includes the *protection* of people against *ionizing radiation* and *safety*; it does not include non-radiation-related aspects of *safety*.

       ⓘ *Protection and safety* is concerned with both *radiation risks* under normal circumstances and *radiation risks* as a consequence of *incidents*, as well as with other possible direct consequences of a loss of *control* over a nuclear reactor core, nuclear chain reaction, *radioactive source* or any other *source* of *radiation*.

       ⓘ *Safety measures* include actions to prevent *incidents* and arrangements put in place to mitigate their consequences if they were to occur.

protection of the environment

Protection and conservation of: non-human species, both animal and plant, and their biodiversity; environmental goods and services such as the production of *food* and *feed*; resources used in agriculture, forestry, fisheries and tourism; amenities used in spiritual, cultural and recreational activities; media such as soil, water and air; and natural processes such as carbon, nitrogen and water cycles.

ⓘ See also *environment*.

protection quantities

Dose quantities developed for purposes of *radiological protection* that allow quantification of the extent of *exposure* of the human body to *ionizing radiation* due to both whole body and partial body external irradiation and *intakes* of radionuclides.

ⓘ Dosimetric quantities that are designated as *protection quantities* are intended for specifying and calculating the numerical *limits* and *levels* that are used in *safety standards* for *radiation protection*.

ⓘ *Protection quantities* relate the magnitude of *exposures* to the *risks* of *health effects of radiation* in a way that is applicable to an individual and that is largely independent of the type of *radiation* and the nature of the *exposure* (internal or external).

ⓘ *Protection quantities* were developed to provide an index of the *risks* arising from the energy imparted by *radiation* to tissue.

protection system

See *plant equipment (for a nuclear power plant)*.

protective action

1.   An action for the purposes of avoiding or reducing *doses* that might otherwise be received in an *emergency exposure situation* or an *existing exposure situation*.

See also *remedial action*.

ⓘ This is related to *radiation protection* (see *protection* (1), and *protection and safety*).

early protective action. A *protective action* in the event of a *nuclear or radiological emergency* that can be implemented within days to weeks and can still be effective.

ⓘ The most common early *protective actions* are relocation and longer term restriction of the consumption of *food* potentially affected by *contamination*.

mitigatory action. Immediate action by the *operator* or other party:

(1)   To reduce the potential for conditions to develop that would result in *exposure* or a release of radioactive material requiring *emergency response actions* on the site or off the site; or

(2)   To mitigate *source* conditions that may result in *exposure* or a *release of radioactive material* requiring *emergency response actions* on the site or off the site.

urgent protective action. A *protective action* in the event of a *nuclear or radiological emergency* which must be taken promptly (usually within hours to a day) in order to be effective, and the effectiveness of which will be markedly reduced if it is delayed.

ⓘ *Urgent protective actions* include *iodine thyroid blocking, evacuation,* short term *sheltering,* actions to reduce inadvertent ingestion, *decontamination* of individuals and prevention of ingestion of *food,* milk and drinking water possibly with *contamination.*

ⓘ A ***precautionary urgent protective action*** is an *urgent protective action* taken before or shortly after a release of *radioactive material,* or an *exposure,* on the basis of the prevailing conditions to avoid or to minimize *severe deterministic effects.*

2.   A *protection system* action calling for the *operation* of a particular safety *actuation device.*

ⓘ This is related to definition (2) of *protection.*

protective task

The generation of at least those *protective actions* necessary to ensure that the *safety task* required by a given *initiating event* is accomplished.

ⓘ This is related to definition (2) of *protection.*

public exposure

See *exposure categories.*

publication, IAEA

See *IAEA publication.*

Q

qualification

Process of determining whether a *system* or *component* is suitable for operational use.

ⓘ *Qualification* is generally performed in the context of a specific set of *qualification requirements* for the specific *facility* and class of *system* and for the specific application.

ⓘ *Qualification* may be accomplished in stages: for example, first, by the qualification of pre-existing equipment (usually early in the *system* realization *process*), then, in a second step, by the *qualification* of the integrated *system* (i.e. in the final realized *design*).

ⓘ *Qualification* may rely on *activities* performed outside the framework of a specific *facility design* (this is called 'generic *qualification*' or 'prequalification').

ⓘ Prequalification may significantly reduce the necessary effort in *facility* specific *qualification*; however, the application specific *qualification requirements* must still be met and be shown to be met.

equipment qualification. Generation and *maintenance* of evidence to ensure that equipment will operate on demand, under *specified service conditions*, to meet *system* performance *requirements*.

See also GSR Part 4 (Rev. 1) [19].

ⓘ More specific terms are used for particular equipment or particular conditions; for example, **seismic qualification** is a form of *equipment qualification* that relates to conditions that could be encountered in the event of earthquakes.

ⓘ The proof that an item of equipment can perform its function, which is an important part of *equipment qualification*, is sometimes termed **substantiation**.

See also harsh environment, mild environment.

qualified equipment

Equipment certified as having satisfied *equipment qualification requirements* for the conditions relevant to its *safety function(s)*.

See also mission time.

qualified expert

An individual who, by virtue of *certification* by appropriate boards or societies, professional licence or academic qualifications and experience, is duly recognized as having expertise in a relevant field of specialization, for example medical physics, *radiation protection*, occupational health, fire safety, quality management or any relevant engineering or *safety* speciality.

qualified life

See *life, lifetime*.

quality assurance

The function of a *management system* that provides confidence that specified *requirements* will be fulfilled.

> ! The IAEA revised the *requirements* and guidance in the subject area of *quality assurance* for its *safety standards* on *management systems* for the *safety* of *facilities and activities* involving the use of *ionizing radiation*.

> ! The terms quality management and *management system* were adopted in the revised standards in place of the terms *quality assurance* and *quality assurance* programme.

> ⓘ Planned and systematic actions are necessary to provide adequate confidence that an item, *process* or service will satisfy given *requirements* for quality; for example, those specified in the *licence*.

> ⓘ This statement was slightly modified from that in the International Organization for Standardization's publication ISO 921:1997 [17] to say 'an item, process or service' instead of 'a product or service' and to add the example.

> ⓘ A more general definition of *quality assurance* (all those planned and systematic actions necessary to provide confidence that a *structure, system or component* will perform satisfactorily in service) and definitions of related terms can be found in the International Organization for Standardization's publication ISO 9000:2015 [46].

quality control (QC)

Part of quality management intended to verify that *structures, systems and components* correspond to predetermined *requirements*.

> ⓘ This definition is taken from ISO 921:1997 (Nuclear Energy: Vocabulary) [17]. A more general definition of *quality control* and definitions of related terms can be found in ISO 9000:2015 [46].

quality factor, Q

A number by which the *absorbed dose* in a tissue or organ is multiplied to reflect the *relative biological effectiveness* of the *radiation*, the result being the *dose equivalent*.

> ⓘ Superseded by *radiation weighting factor* in the definition of *equivalent dose* in Ref. [44], but still defined, as a function of *linear energy transfer*, for use in calculating the *dose equivalent quantities* used in *monitoring*.

> ⓘ GSR Part 3 [1] also states that the mean *quality factor* $\overline{Q}$ at 10 mm depth in the *ICRU sphere* can be used as a value of *radiation weighting factor* for *radiation* types for which GSR Part 3 does not specify a value (see *radiation weighting factor*).

R

[rad]

Unit of *absorbed dose*, equal to 0.01 Gy.

ⓘ Superseded by the *gray* (Gy).

ⓘ Abbreviation of *röntgen absorbed dose* or *radiation absorbed dose*.

radiation

! When used in *IAEA publications*, the term *radiation* usually refers to *ionizing radiation* only. The IAEA has no statutory responsibilities in relation to non-ionizing *radiation*.

ⓘ *Ionizing radiation* can be divided into *low linear energy transfer radiation* and *high linear energy transfer radiation* (as a guide to its *relative biological effectiveness*), or into *strongly penetrating radiation* and *weakly penetrating radiation* (as an indication of its ability to penetrate shielding or the human body).

high linear energy transfer (LET) radiation. *Radiation* with high *linear energy transfer*, normally assumed to comprise protons, neutrons and alpha particles (or other particles of similar or greater mass).

ⓘ These are the types of *radiation* for which the International Commission on Radiological Protection recommends a *radiation weighting factor* greater than 1.

ⓘ Contrasting term: *low linear energy transfer radiation*.

ionizing radiation. For the purposes of *radiation protection*, *radiation* capable of producing ion pairs in biological material(s).

low linear energy transfer (LET) radiation. *Radiation* with low *linear energy transfer*, normally assumed to comprise photons (including X rays and gamma *radiation*), electrons, positrons and muons.

ⓘ These are the types of *radiation* for which the International Commission on Radiological Protection recommends a *radiation weighting factor* of 1.

ⓘ Contrasting term: *high linear energy transfer radiation*.

strongly penetrating radiation. *Radiation* for which *limits* on *effective dose* are generally more restrictive than *limits* on *equivalent dose* to any tissue or organ; that is, the fraction of the relevant *dose limit* received will, for a given *exposure*, be higher for *effective dose* than for *equivalent dose* to any tissue or organ.

ⓘ For most practical purposes, it may be assumed that *strongly penetrating radiation* includes photons of energy above about 12 keV, electrons of energy more than about 2 MeV and neutrons.

ⓘ Contrasting term: *weakly penetrating radiation*.

weakly penetrating radiation. *Radiation* for which *limits* on *equivalent dose* to any tissue or organ are generally more restrictive than *limits* on *effective dose*; that is, the fraction of the relevant *dose limit* received will, for a given *exposure*, be higher for *equivalent dose* to any tissue or organ than for *effective dose*.

ⓘ For most practical purposes, it may be assumed that *weakly penetrating radiation* includes photons of energy below about 12 keV, electrons of energy less than about 2 MeV, and massive charged particles such as protons and alpha particles.

ⓘ Contrasting term: *strongly penetrating radiation.*

[radiation area]

See *area: controlled area* and *supervised area.*

radiation detriment

The total harm that would eventually be incurred by a group that is subject to *exposure* and by its descendants as a result of the group's *exposure* to *radiation* from a *source.*

ⓘ In its Publication 60 [44], the International Commission on Radiological Protection defines a measure of *radiation detriment* that has the dimensions of probability, and that could therefore also be considered a measure of *risk.*

radiation emergency

See *emergency: nuclear or radiological emergency.*

radiation exposure device

See device.

radiation generator

See *source* (1).

[radiation level]

The corresponding *dose rate* expressed in millisieverts per hour or microsieverts per hour.

! This usage was specific to previous editions of the Transport Regulations [2] and should be avoided.

radiation protection

See *protection* (1).

radiation protection officer

A person technically competent in *radiation protection* matters relevant for a given type of *practice* who is designated by the *registrant, licensee* or *employer* to oversee the application of regulatory *requirements.*

radiation protection programme

Systematic arrangements that are aimed at providing adequate consideration of *radiation protection* measures. (See SSR-6 (Rev. 1) [2].)

radiation risks

Detrimental *health effects* of *exposure* to *radiation* (including the likelihood of such effects occurring), and any other *safety* related *risks* (including those to the *environment*) that might arise as a direct consequence of:

(a) *Exposure* to *radiation*;

(b) The presence of *radioactive material* (including *radioactive waste*) or its *release* to the *environment*;

(c) A loss of *control* over a nuclear reactor core, nuclear chain reaction, *radioactive source* or any other *source* of *radiation*. (See SF-1 [24].)

    ⓘ For the purposes of the IAEA *safety standards*, it is assumed that there is no threshold level of *radiation dose* below which there are no associated *radiation risks*.

    ⓘ Safety Requirements and Safety Guides specify the *radiation exposures* and other *radiation risks* to which they refer.

radiation safety

See *safety* and protection and safety.

radiation search

The set of activities to detect and identify suspicious nuclear or other *radioactive material out of regulatory control* and to determine its location.

    ! This usage is specific to the area of *nuclear security*, in particular publications addressing *radioactive material out of regulatory control*, and should otherwise be avoided.

radiation source

See *source* (1).

radiation survey

Activities to map the radiation background of natural and human made *radioactive material* in an area or to facilitate later search activities.

    ! This usage is specific to the area of *nuclear security*, in particular publications addressing *radioactive material out of regulatory control*, and should otherwise be avoided.

radiation weighting factor, w_R

A number by which the *absorbed dose* in a tissue or organ is multiplied to reflect the *relative biological effectiveness* of the *radiation* in inducing *stochastic effects* at low *doses*, the result being the *equivalent dose*.

    ⓘ Values are selected by the International Commission on Radiological Protection to be representative of the relevant *relative biological effectiveness* and are broadly compatible with the values previously recommended for *quality factors* in the definition of *dose equivalent*.

    ⓘ The *radiation weighting factor* values recommended in Ref. [33] are set out in Table 3.

TABLE 3. RADIATION WEIGHTING FACTORS RECOMMENDED IN REF. [33]

Type of *radiation*	w_R
Photons, all energies	1
Electrons and muons, all energies[a]	1
Protons and charged pions	2
Alpha particles, *fission fragments*, heavy ions	20
Neutrons	A continuous function of neutron energy:

$$w_R = \begin{cases} 2.5 + 18.2\,e^{-[\ln(E_n)]^2/6}, & E_n < 1\ MeV \\ 5.0 + 17.0\,e^{-[\ln(2E_n)]^2/6}, & 1\ MeV \leq E_n \leq 50\ MeV \\ 2.5 + 3.25\,e^{-[\ln(0.04E_n)]^2/6}, & E_n > 50\ MeV \end{cases}$$

Note: All values relate to the radiation incident on the body or, for internal radiation sources, radiation emitted from the incorporated radionuclide(s).

[a] Excluding Auger electrons emitted from radionuclides bound to deoxyribonucleic acid (DNA) in the human body, for which special microdosimetric considerations apply.

ⓘ For *radiation* types and energies not included in Table 3, w_R can be taken to be equal to $\overline{Q}$ at 10 mm depth in the *ICRU sphere* and can be obtained as follows:

$$\overline{Q} = \frac{1}{D}\int_0^\infty Q(L)D_L\,\mathrm{d}L$$

where D is the *absorbed dose*, $Q(L)$ is the *quality factor* in terms of the *unrestricted linear energy transfer L* in water, specified in Ref. [36], and D_L is the distribution of D in L.

$$Q(L) = \begin{cases} 1 & for & L \leq 10 \\ 0.32L - 2.2 & for & 10 < L < 100 \\ 300/\sqrt{L} & for & L \geq 100 \end{cases}$$

where L is expressed in keV/μm.

radioactive

1.   Exhibiting *radioactivity*; emitting or relating to the emission of *ionizing radiation* or particles. (adjective)

    ! This is the 'scientific' definition, and should not be confused with the 'regulatory' definition (2).

2.   Designated in national law or by a *regulatory body* as being subject to *regulatory control* because of its *radioactivity*. (adjective)

    ! This is the 'regulatory' definition, and should not be confused with the 'scientific' definition (1).

radioactive contents

The *radioactive material* together with any contaminated or activated solids, liquids and gases within the *packaging*. (See SSR-6 (Rev. 1) [2].)

radioactive discharges

See *discharge* (1).

radioactive equilibrium

The state of a *radioactive* decay chain (or part thereof) where the *activity* of each radionuclide in the chain (or part of the chain) is the same.

ⓘ This state is achieved when the parent nuclide has a much longer *half-life* than any of the decay products, and after a time equal to several times the *half-life* of the longest lived of the decay products.

ⓘ Hence, the term 'secular equilibrium' is also used (with secular in this context meaning 'eventual', as contrasted with 'transient equilibrium').

radioactive half-life

See *half-life* (2).

radioactive material

1.   Material designated in national law or by a *regulatory body* as being subject to *regulatory control* because of its *radioactivity*.

ⓘ Some publications in the Nuclear Security Series extend the definition, for the avoidance of doubt, to indicate that "In the absence of such a designation by a State, any material for which protection is required by the current version of the International Basic Safety Standards" [9].

! This is the 'regulatory' meaning of *radioactive* (2), and should not be confused with the 'scientific' meaning of *radioactive* (1): 'exhibiting *radioactivity*; emitting or relating to the emission of *ionizing radiation* or particles'.

! The 'scientific' meaning of *radioactive* (1) — as in **radioactive substance** — refers only to the presence of *radioactivity*, and gives no indication of the magnitude of the *hazard* involved.

! The term *radioactive substance* is also used to indicate that the 'scientific' meaning of *radioactive* (see *radioactive* (1)) is intended, rather than the 'regulatory' meaning of *radioactive* (see *radioactive* (2)) suggested by the term *radioactive material*.

! However, in some States the term *radioactive substance* is used for the 'regulatory' purpose. It is therefore essential that any such distinctions in meaning are clarified.

! In some Nuclear Security Series publications (e.g. Refs [6, 7]) it is stated or implied that *radioactive substance* is synonymous with *radioactive material*, in order to confirm consistency with the 2005 CPPNM Amendment [5]. In contexts related to the CPPNM Amendment, the terms are synonymous, but in general they may have different meanings.

ⓘ In regulatory terminology in some States, *radioactive material* ceases to be *radioactive material* when it becomes *radioactive waste*; the term *radioactive substance* is used to cover both, that is *radioactive substance* includes *radioactive material* and *radioactive waste*.

ⓘ *Radioactive material* should be used in the singular unless reference is expressly being made to the presence of various types of *radioactive material*.

2.   Any material containing radionuclides where both the *activity concentration* and the total *activity* in the *consignment* exceed the values specified in [paras 402 – 407 of the Transport Regulations [2]]. (See SSR-6 (Rev. 1) [2].)

! This usage is specific to the Transport Regulations [2], and should otherwise be avoided.

radioactive source

See *source* (2).

radioactive sources, safety of

See *safety of radioactive sources*.

radioactive substance

See *radioactive material* (1).

radioactive waste

1.   For legal and regulatory purposes, material for which no further use is foreseen that contains, or is contaminated with, radionuclides at *activity concentrations* greater than *clearance levels* as established by the *regulatory body*.

   ⓘ In effect, *radioactive material* in gaseous, liquid or solid form for which no further use is foreseen.

   ! It should be recognized that this definition is purely for regulatory purposes, and that material with *activity concentrations* equal to or less than *clearance levels* is *radioactive* from a physical viewpoint, although the associated radiological *hazards* are considered negligible.

   See also *radioactive, radioactive material* and *radioactive substance*.

   ⓘ *Waste* should be used in the singular unless reference is expressly being made to the presence of various types of *waste*.

2.   [*Radioactive material* in gaseous, liquid or solid form for which no further use is foreseen by the Contracting Party or by a natural or *legal person* whose decision is accepted by the Contracting Party, and which is controlled as *radioactive waste* by a *regulatory body* under the legislative and regulatory framework of the Contracting Party.] (See Ref. [11].)

radioactive waste management

1.   All administrative and operational *activities* involved in the handling, *pretreatment, treatment, conditioning, transport, storage* and *disposal* of *radioactive waste*.

   conditioning. Those *operations* that produce a *waste package* suitable for handling, *transport, storage* and/or *disposal*.

   ⓘ *Conditioning* may include the conversion of the *waste* to a solid *waste form*, enclosure of the *waste* in containers and, if necessary, provision of an *overpack*.

   immobilization. Conversion of *waste* into a *waste form* by solidification, embedding or encapsulation.

   ⓘ *Immobilization* reduces the potential for *migration* or *dispersion* of radionuclides during handling, *transport, storage* and/or *disposal*.

   overpack. A secondary (or additional) outer container for one or more *waste packages*, used for handling, *transport, storage* and/or *disposal*.

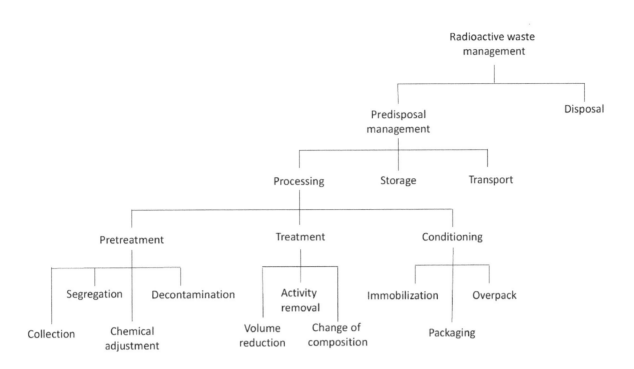

FIG. 5. Activities involved in radioactive waste management.

packaging. Preparation of *radioactive waste* for safe handling, *transport, storage* and/or *disposal* by means of enclosing it in a suitable *container*.

predisposal management. Any *waste management* steps carried out prior to *disposal*, such as *pretreatment, treatment, conditioning, storage* and *transport activities*.

ⓘ Predisposal is not a form of *disposal*: predisposal is used as a contraction of '*pre-disposal management* of *radioactive waste*'.

pretreatment. Any or all of the *operations* prior to *waste treatment*, such as collection, *segregation*, chemical adjustment and *decontamination*.

processing. Any *operation* that changes the characteristics of *waste*, including *pretreatment, treatment* and *conditioning*.

segregation. An activity where types of *waste* or material (*radioactive* or exempt) are separated or are kept separate on the basis of radiological, chemical and/or physical properties, to facilitate *waste* handling and/or *processing*.

treatment. *Operations* intended to benefit *safety* and/or economy by changing the characteristics of the *waste*. Three basic *treatment* objectives are:

(a) *Volume reduction*;

(b) Removal of radionuclides from the *waste*;

(c) Change of composition.

Treatment may result in an appropriate *waste form*.

ⓘ If *treatment* does not result in an appropriate *waste form*, the *waste* may be immobilized.

volume reduction. A *treatment* method that decreases the physical volume of a *waste*.

ⓘ Typical *volume reduction* methods are mechanical compaction, incineration and evaporation.

ⓘ Should not be confused with *waste minimization*.

See also *minimization of waste*.

2.   [All *activities*, including *decommissioning activities*, that relate to the handling, *pretreatment*, *treatment*, *conditioning*, *storage* or *disposal* of *radioactive waste*, excluding *off-site* transportation. It may also involve *discharges*.] (See Ref. [11].)

radioactive waste management facility

1.   *Facility* specifically designed to handle, treat, condition, store or permanently dispose of *radioactive waste*.

2.   [Any *facility* or installation the primary purpose of which is *radioactive waste management*, including a nuclear facility in the process of being decommissioned only if it is designated by the Contracting Party as a *radioactive waste management facility*.] (See Ref. [11].)

radioactivity

The phenomenon whereby atoms undergo spontaneous random disintegration, usually accompanied by the emission of *radiation*.

! In *IAEA publications*, *radioactivity* should be used only to refer to the phenomenon.

! To refer to the physical quantity or to an amount of a *radioactive substance*, use *activity*.

radiochronometry

The use of measurements of radioactive decay products in a sample of material to determine the time elapsed since the last separation of progeny from the parent material (and thus, the 'age' of the material in the measured sample).

radiological assessor

1.   A person or team who in the event of a *nuclear or radiological emergency* assists the *operator* or *off-site response organizations* by performing radiological surveys, performing *dose assessments*, controlling *contamination*, ensuring the *radiation protection* of *emergency workers* and formulating recommendations on *protective actions* and *other response actions*.

ⓘ The *radiological assessor* could be the *radiation protection officer*.

2.   A person who, at a *radiological crime scene*, assists by performing radiation surveys, performing dose assessments, assisting with the control of radionuclide contamination, ensuring the radiation protection of *crime scene* personnel and formulating recommendations on protective actions.

! This usage is specific to the area of *nuclear security*, in particular publications addressing *radioactive material out of regulatory control*, and should otherwise be avoided.

radiological crime scene

See *crime scene*.

radiological dispersal device

See *device*.

radiological emergency

See *emergency*.

radiological environmental impact assessment

See *assessment* (1).

[radiological material]

    ! Avoid this term.

See also *nuclear material* and *radioactive material*.

radiological medical practitioner

A *health professional* with specialist education and training in the medical uses of radiation, who is competent to perform independently or to oversee *radiological procedures* in a given specialty.

    ⓘ Competence of persons is normally assessed by the State by having a formal mechanism for registration, accreditation or *certification* of *radiological medical practitioners* in the given specialty (e.g. radiology, radiation therapy, nuclear medicine, dentistry, cardiology).

    ⓘ States that have yet to develop such a mechanism need to assess the education, training and competence of any individual proposed by the *licensee* to act as a *radiological medical practitioner* and to decide, on the basis of either international standards or standards of a State where such a system exists, whether such an individual could undertake the functions of a *radiological medical practitioner*, within the required specialty.

radiological procedure

A medical imaging *procedure* or therapeutic *procedure* that involves *ionizing radiation* — such as a *procedure* in diagnostic radiology, nuclear medicine or radiation therapy, or a planning *procedure*, image guided interventional *procedure* or other interventional *procedure* involving *radiation* — delivered by a *radiation generator*, a device containing a *sealed source* or an *unsealed source*, or by means of a radiopharmaceutical administered to a *patient*.

radiological protection

See *protection* (1).

[radionuclear]

! 'Radionuclear' is not a legitimate word.

See also *nuclear material* and *radioactive material*.

! *Radionuclear* has been used in nuclear medicine to mean 'involving the use of radionuclides'; thus '*radionuclear* tests' has been used in nuclear medicine to mean tests in which radiopharmaceuticals are administered. This usage is to be avoided.

! *Radionuclear* has also been used as a journalese shorthand form for 'nuclear and/or radiological', as in the terms '*radionuclear* weapon' and '*radionuclear* emergency'; or used for 'nuclear and/or *radioactive*', as in the term '*radionuclear* material'. These and other such usages are to be avoided.

radionuclides of artificial origin

See *radionuclides of natural origin*.

radionuclides of natural origin

Radionuclides that occur naturally on Earth in significant quantities.

ⓘ The term is usually used to refer to the primordial radionuclides ^{40}K, ^{235}U, ^{238}U, ^{232}Th and their *radioactive* decay products.

ⓘ Contrasted with **radionuclides of artificial origin**, anthropogenic radionuclides and human made radionuclides (which all mean the same), and also with artificial radionuclides (which exclude *radionuclides of artificial origin* that are also naturally occurring).

! *Radionuclides of artificial origin* may include radionuclides that are also naturally occurring but may not include *radionuclides of natural origin*.

radiopharmacist

A *health professional*, with specialist education and training in radiopharmacy, who is competent to prepare and dispense radiopharmaceuticals used for the purposes of medical diagnosis and radionuclide therapy.

ⓘ Competence of persons is normally assessed by the State by having a formal mechanism for registration, accreditation or *certification* of radiopharmacists.

ⓘ States that have yet to develop such a mechanism need to assess the education, training and competence of any individual proposed by the *licensee* to act as a radiopharmacist and to decide, on the basis of either international standards or standards of a State where such a system exists, whether such an individual could undertake the functions of a radiopharmacist.

radon

1.   Any combination of isotopes of the element *radon*.

ⓘ For the purposes of the IAEA *safety standards*, radon refers to ^{220}Rn and ^{222}Rn.

2.   [^{222}Rn.]

ⓘ When contrasted with *thoron* (^{220}Rn).

radon progeny

The short lived *radioactive* decay products of ^{220}Rn and of ^{222}Rn.

ⓘ For ^{222}Rn, this includes the decay chain up to but not including ^{210}Pb, namely ^{218}Po, ^{214}Pb, ^{214}Bi and ^{214}Po, plus traces of ^{218}At, ^{210}Tl and ^{209}Pb. Lead-210, which has a *half-life* of 22.3 years, and its *radioactive* progeny — ^{210}Bi and ^{210}Po, plus traces of ^{206}Hg and ^{206}Tl — are, strictly, progeny of ^{222}Rn, but they are not included in this listing because they will not normally be present in significant amounts in airborne form. For ^{220}Rn, this includes ^{216}Po, ^{212}Pb, ^{212}Bi, ^{212}Po and ^{208}Tl.

reactivity, ρ

For a nuclear chain reacting medium:

$$\rho = 1 - \frac{1}{k_{\text{eff}}}$$

where k_{eff} is the ratio between the number of fissions in two succeeding generations (later to earlier) of the chain reaction.

ⓘ A measure of the *deviation* from *criticality* of a nuclear chain reacting medium, such that positive values correspond to a supercritical state and negative values correspond to a subcritical state.

shutdown reactivity. The *reactivity* when all *control* devices are introducing their maximum negative *reactivity*.

ⓘ A reactor is shut down quickly by moving *control* devices rapidly into position to introduce their negative *reactivity* into the reactor core.

receiver

See *consignee*.

recording level

See *level*.

recycling

See *minimization of waste*.

redundancy

Provision of alternative (identical or diverse) *structures, systems and components*, so that any single *structure, system or component* can perform the required function regardless of the state of *operation* or *failure* of any other.

reference air kerma rate

See *kerma*.

reference individual

An idealized human with characteristics defined by the International Commission on Radiological Protection for *radiation protection* purposes.

ⓘ Reference values for eight *reference individuals* — a newborn; a one year old; a five year old; a ten year old; male and female 15 year olds; and male and female adults — are given in Ref. [50].

ⓘ These reference values are based on data for western European and North American populations, but Ref. [50] also provides additional information on individual variation among grossly normal individuals resulting from differences in age, gender, ethnicity and other factors.

ⓘ This is a refinement of the *Reference Man* concept.

reference level

See *level*.

[Reference Man]

An idealized adult Caucasian human male defined by the International Commission on Radiological Protection for the purpose of *radiation protection assessment*.

See Ref. [51].

ⓘ Although *Reference Man* is now being superseded by the more general concept of the *reference individual* (see Refs [50, 52]), some concepts and quantities are still defined in terms of *Reference Man*.

reference scenario

See *scenario*.

referring medical practitioner

A *health professional* who, in accordance with national *requirements*, may refer individuals to a *radiological medical practitioner* for *medical exposure*.

registrant

The holder of a current *registration*.

registration

A form of *authorization* for *facilities and activities* of low or moderate *risks* whereby the *person or organization* responsible for the *practice* has, as appropriate, prepared and submitted a *safety assessment* of the *facilities* and equipment to the *regulatory body*. The *practice* or use is authorized with conditions or limitations as appropriate.

ⓘ The *requirements* for *safety assessment* and the conditions or limitations applied to the *facilities and activities* would be less severe for *registration* than those for issuing a *licence*.

ⓘ Typical *facilities and activities* that are amenable to *registration* are those for which: (a) *safety* can largely be ensured by the *design* of the *facilities* and equipment; (b) the operating *procedures* are simple to follow; (c) the *safety* training *requirements* are minimal; and (d) there is a history of few problems with *safety* in *operations*. *Registration* is best suited to those *facilities and activities* for which *operations* do not vary significantly.

See also *registrant*.

ⓘ Derivative terms should not be needed; a *registration* is a product of the *authorization process*, and a *facility or activity* with a current *registration* is an *authorized facility or activity*.

regulatory authority

1.   See *regulatory body* (2).

2.   [An authority or authorities designated or otherwise recognized by a government for regulatory purposes in connection with *protection and safety*.]

! Superseded by the term *regulatory body*, which should be used.

regulatory body

1.   An authority or a system of authorities designated by the government of a State as having legal authority for conducting the regulatory *process*, including issuing *authorizations*, and thereby regulating the *nuclear*, *radiation*, *radioactive waste* and *transport safety*.

ⓘ The *regulatory body* is generally a national entity, established and empowered by law, whose organization, management, functions, processes, responsibilities and competences are subject to the *requirements* of IAEA *safety standards*.

ⓘ The national *competent authority* for the regulation of *radioactive material transport safety* (see SSR-6 (Rev. 1) [2]) is included in this description, as is the *regulatory body* for *protection and safety*.

! Supersedes the term *regulatory authority*, which should not be used.

2.   One or more authorities designated by the government of a State as having legal authority for conducting the regulatory process, including issuing authorizations.

ⓘ The term *regulatory authority* may also be used with this definition to avoid confusion in *nuclear security* contexts in which the term *regulatory body* might be assumed by readers to imply only the *regulatory body* for *safety*. The term "*competent authority* with regulatory responsibility" is also used in Ref. [8] for this purpose.

3.   [For each Contracting Party any body or bodies given the legal authority by that Contracting Party to grant *licences* and to regulate the *siting, design, construction, commissioning, operation* or *decommissioning* of *nuclear installations*.] (See Ref. [10].)

4.   [Any body or bodies given the legal authority by the Contracting Party to regulate any aspect of the *safety* of *spent fuel* or *radioactive waste management* including the granting of *licences*.] (See Ref. [11].)

5.   [An entity or organization or a system of entities or organizations designated by the government of a State as having legal authority for exercising *regulatory control* with respect to *radioactive sources*, including issuing *authorizations*, and thereby regulating one or more aspects of the *safety* or *security* of *radioactive sources*.] (See Ref. [21].)

regulatory control

See *control* (1).

regulatory inspection

See *inspection*.

relative biological effectiveness (RBE)

A measure of the relative effectiveness of different *radiation* types at inducing a specified *health effect*, expressed as the inverse ratio of the *absorbed doses* of two different *radiation* types that would produce the same degree of a defined biological *end point*.

ⓘ Values of *relative biological effectiveness* in causing the development of *deterministic effects* are selected to be representative of the *severe deterministic effects* that are significant to *emergency preparedness* and *response*.

ⓘ The tissue or organ specific and radiation type specific values of $RBE_{T,R}$ for the development of selected *severe deterministic effects* are as shown in Table 4.

TABLE 4. VALUES OF RELATIVE BIOLOGICAL EFFECTIVENESS

Health effect	Critical tissue or organ	Exposure[a]	$RBE_{T,R}$
Haematopoietic syndrome	Red marrow	External and internal γ	1
		External and internal n	3
		Internal β	1
		Internal α	2
Pneumonitis	Lung[b]	External and internal γ	1
		External and internal n	3
		Internal β	1
		Internal α	7
Gastrointestinal syndrome	Colon	External and internal γ	1
		External and internal n	3
		Internal β	1
		Internal α	0[c]
Necrosis	Tissue[d]	External β, γ	1
		External n	3
Moist desquamation	Skin[e]	External β, γ	1
		External n	3
Hypothyroidism	Thyroid	Intake of iodine isotopes[f]	0.2
		Other thyroid seekers	1

[a] External β, γ exposure includes exposure due to bremsstrahlung produced within the material of the source.

[b] Tissue of the alveolar–interstitial region of the respiratory tract.

[c] For alpha emitters uniformly distributed in the contents of the colon, it is assumed that irradiation of the walls of the intestine is negligible.

[d] Tissue at a depth of 5 mm below the skin surface over an area of more than 100 cm².

[e] Tissue at a depth of 0.4 mm below the skin surface over an area of more than 100 cm².

Uniform irradiation of the tissue of the thyroid gland is considered to be five times more likely to produce *deterministic effects* than internal exposure due to low energy beta emitting isotopes of iodine such as ^{131}I, [129]I, ^{125}I, [124]I and [123]I. Thyroid seeking radionuclides have a heterogeneous distribution in thyroid tissue. The isotope ^{131}I emits low energy beta particles, which leads to a reduced effectiveness of irradiation of critical thyroid tissue owing to the dissipation of the energy of the particles within other tissues.

relative biological effectiveness (RBE) weighted absorbed dose, AD$_T$

See *dose quantities*: *absorbed dose*

relative risk

See *risk* (3).

release

The action or process of setting free or being set free, or of allowing or being allowed to move or flow freely.

> ! *Release* is used in both a physical 'scientific' sense (see *discharge* (1)) and a 'regulatory' sense (see *clearance*), as well as in the usual sense of, for example, a release of energy.

reliability

The probability that a *system* or *component* or an item will meet its minimum performance *requirements* when called upon to do so, for a specified period of time and under stated *operating conditions*.

> ⓘ The *reliability* of a computer based *system*, for example, includes the *reliability* of hardware, which is usually quantified, and the reliability of software, which is usually a qualitative measure as there are no generally recognized means of quantifying the *reliability* of software.

See also *availability*.

reliability centred maintenance

See *maintenance*.

relocation

The non-urgent removal or extended exclusion of people from an area to avoid long term *exposure* from deposited *radioactive* material.

> ⓘ *Relocation* is an *early protective action*. It may be a substitution for the *urgent protective action* of *evacuation*.

See also *evacuation*.

> ⓘ *Relocation* is considered to be **permanent relocation** if return is not foreseeable; otherwise it is **temporary relocation.**

[rem]

Unit of *dose equivalent* and *effective dose equivalent*, equal to 0.01 Sv.

    ① Superseded by the *sievert (Sv)*.

    ① Abbreviation of *röntgen* equivalent man.

remedial action

The removal of a *source* or the reduction of its magnitude (in terms of *activity* or amount) for the purposes of preventing or reducing *exposures* that might otherwise occur in an *emergency* or in an *existing exposure situation*.

    ① *Remedial actions* could also be termed *protective actions*, but *protective actions* are not necessarily *remedial actions*.

See also *remediation* and *protective action*.

remediation

Any measures that may be carried out to reduce the *radiation exposure* due to existing *contamination* of land areas through actions applied to the *contamination* itself (the *source*) or to the *exposure pathways* to humans.

    ① Complete removal of the *contamination* is not implied.

    ① The use of the terms *cleanup*, rehabilitation and restoration as synonyms for *remediation* is discouraged. Such terms may be taken to imply that the conditions that prevailed before the *contamination* can be achieved again and unconditional use of the land areas can be restored, which is not usually the case (e.g. owing to the effects of the *remedial action* itself). Often *remediation* is used to restore land areas to conditions suitable for limited use under *institutional control*.

    ① *Remediation* can entail activities that are similar to *decommissioning*; both *remediation* and *decommissioning* activities are typically performed under an *authorization*. Abandoned and presently unauthorized industrial sites, such as former uranium mines and mills and former radium processing facilities, may have buildings and structures that are taken down by actions consistent with the *decommissioning* process (e.g. *decontamination* and *dismantling*); however, such *activities* are considered to be a part of site *remediation*.

    ① In some contexts (e.g. the wider chemical industry), the terms remediation and restoration are used to describe different parts of overall recovery.

    ① The term *cleanup* is used in the context of *decommissioning*.

See also *decontamination*.

remediation plan

A document setting out the various *activities* and actions and the timescales necessary to apply the approach and to achieve the objectives of the remediation strategy in order to meet the legal and regulatory *requirements* for *remediation*.

remedy

Corrective actions designed to address *root causes*.

See also *cause*: *root cause*.

removable contamination

See *contamination* (2): *non-fixed contamination*

repair

An action on a non-conforming product to make it acceptable for its intended use (see Ref. [46]).

See also *cause*: *direct cause*.

repository

ⓘ Synonymous with *disposal facility*.

representative person

An individual receiving a *dose* that is representative of the *doses* to the more highly exposed individuals in the population.

ⓘ The *representative person* will generally be a hypothetical construct and not an actual member of the population. The concept is used to determine compliance or in prospective *assessments*.

ⓘ In estimating the *dose* to the *representative person*, a number of factors are taken into account for the population exposed: (i) all relevant *exposure pathways* for the *source* and all locations under consideration; (ii) the spatial distribution of radionuclides in the *environment*, to ensure that individuals with higher exposures are included; (iii) age dependent physiological parameters and information on diet, habits, residence and use of local resources; (iv) dosimetric *models* and appropriate *dose coefficients*.

ⓘ Application of the concept of a *representative person* to *potential exposures*, such as those that may occur in the future as a result of *radioactive waste disposal*, is complicated by the facts that both the *dose* (if it occurs) and the probability of receiving the *dose* are relevant, and that these two parameters are essentially independent of one another.

ⓘ Hence, a population can be homogeneous with respect to *dose* but not *risk*, and, more importantly, vice versa.

ⓘ A possible approach is to define a *representative person* that is reasonably representative with respect to *risk*, and that is typical of those people who might be subject to the highest *risk*.

ⓘ ICRP Publication 101 [52] indicates that the *dose* to the *representative person* "is the equivalent of, and replaces, the mean dose in the 'critical group'", and provides guidance on assessing *doses* to the *representative person*. The concept of critical group remains valid.

See also *member of the public*.

representative threat statement

See *threat*

reprocessing

A *process* or *operation*, the purpose of which is to extract *radioactive* isotopes from *spent fuel* for further use.

requirement (safety)

That which is established or required by the Fundamental Safety Principles (IAEA Safety Fundamentals) [24] or IAEA Safety Requirements publications or by (national or international) laws or regulations.

> ! In *IAEA publications*, *requirement* (and 'required' and other words deriving from the verb 'to require') should be used in this sense only. Care should be taken to avoid confusion: the use of 'requirement' in the more general sense of something that is necessary should be avoided.

> ⓘ *Requirements*, including numbered 'overarching' *requirements*, are expressed as 'shall' statements. Reported (quoted) *requirements*, e.g. in a Safety Guide, are reported using a formulation such as 'it is required to…'.

requirements engineering

An engineering process that includes the activities involved in developing, documenting and maintaining a set of requirements.

research reactor

[A nuclear reactor used mainly for the generation and utilization of neutron flux and *ionizing radiation* for research and other purposes, including experimental *facilities* associated with the reactor and *storage*, handling and *treatment facilities* for *radioactive material* on the same site that are directly related to safe *operation* of the *research reactor*. *Facilities* commonly known as *critical assemblies* are included.]

> ! This definition is particular to the Code of Conduct on the Safety of Research Reactors [53].

residual dose

See *dose concepts*.

residual heat

The sum of the heat originating from *radioactive* decay and *shutdown* fission and the heat stored in reactor related *structures* and in heat *transport* media.

response

1.   See *emergency response*.

2.   All of the activities by a State that involve assessing and responding to a *nuclear security event*.

§ In safety, 'response' normally refers to response to a *nuclear or radiological emergency*, i.e. triggered by an accident or a *nuclear security event*. In nuclear security, 'response' normally refers to measures that include activities for the identification, collection, packaging and transport of evidence contaminated with radionuclides, nuclear forensics and related actions in the context of investigation into the circumstances surrounding a nuclear security event.

response forces. Persons, on-site or off-site, who are armed and appropriately equipped and trained to counter an attempted unauthorized removal or an act of sabotage.

response measure. A measure intended to assess an alarm/alert and to respond to a *nuclear security event*.

See also *emergency response action* and *protective action*.

response system. An integrated set of *response measures* including capabilities and resources necessary for assessing the alarms/alerts and response to *a nuclear security event*.

response forces

See *response (2)*.

response measure

See *response (2)*.

response organization

An organization designated or recognized by a State as being responsible for managing or implementing any aspect of an *emergency response*.

> ⓘ This also includes those organizations or services necessary to support the management and/or conduct of an *emergency response*, such as meteorological services.

response spectrum

A curve calculated from an *accelerogram* that gives the value of peak response in terms of the acceleration, velocity or displacement of a damped single-degree-of-freedom linear oscillator (with a given damping ratio) as a function of its natural frequency or period of vibration.

uniform hazard response spectrum. *Response spectrum* with an equal probability of exceedance for each of its spectral ordinates.

in-structure response spectrum. The seismic *response spectrum* at a particular elevation of a building for a given input ground motion.

response system

See *response (2)*.

response time (of a component)

The period of time necessary for a *component* to achieve a specified output state from the time that it receives a signal requiring it to assume that output state.

> ! Note that this is not related to *emergency response* or to *response* to a *nuclear security event*

restricted linear collision stopping power

See *linear energy transfer (LET)*.

restricted use

See *use*.

reuse

See *minimization of waste*.

risk

> ! Depending on the context, the term *risk* may be used to represent a quantitative measure (as, for example, in definitions (1) and (2)) or as a qualitative concept (as often for definitions (3), (4) and (5)).

1.　　A multiattribute quantity expressing *hazard*, danger or chance of harmful or injurious consequences associated with *exposures* or *potential exposures*. It relates to quantities such as the probability that specific deleterious consequences may arise and the magnitude and character of such consequences.

> ⓘ In mathematical terms, this can be expressed generally as a set of triplets, $R = \{\langle S_i | p_i | X_i \langle \}$, where S_i is an identification or description of a *scenario* i, p_i is the probability of that *scenario* and X_i is a measure of the consequence of the *scenario*. The concept of *risk* is sometimes also considered to include uncertainty in the probabilities p_i of the *scenarios*.

2.　　The mathematical mean (expectation value) of an appropriate measure of a specified (usually unwelcome) consequence:

$$R = \sum_i p_i \cdot C_i$$

where p_i is the probability of occurrence of *scenario* or *event* sequence i and C_i is a measure of the consequence of that *scenario* or *event* sequence.

> ⓘ Typical consequence measures C_i include core damage frequency, the estimated number or probability of *health effects*, etc.

> ⓘ If the number of *scenarios* or *event* sequences is large, the summation is replaced by an integral.

> ! The summing of *risks* associated with *scenarios* or *event* sequences with widely differing values of C_i is controversial. In such cases the use of the term 'expectation value', although mathematically correct, is misleading and should be avoided if possible.

> ⓘ Methods for treating uncertainty in the values of p_i and C_i — and in particular whether such uncertainty is represented as an element of *risk* itself or as uncertainty in estimates of *risk* — vary.

3.　　The probability of a specified *health effect* occurring in a person or group as a result of *exposure* to *radiation*.

> ⓘ The *health effect(s)* in question must be stated — for example, *risk* of fatal cancer, *risk* of serious *hereditary effects* or overall *radiation detriment* — as there is no generally accepted 'default'.

> ⓘ Commonly expressed as the product of the probability that *exposure* will occur and the probability that the *exposure*, assuming that it occurs, will cause the specified *health effect*.

> ⓘ The latter probability is sometimes termed the **conditional risk**.

annual risk. The probability that a specified *health effect* will occur at some time in the future in an individual as a result of *dose* received or *dose* committed in a given year, taking account of the probability of *exposure* occurring in that year.

> ! This is not the probability of the *health effect* occurring in the year in question; it is the *lifetime risk* resulting from the *annual dose* for that year.

attributable risk. The *risk* of a specified *health effect* assumed to result from a specified *exposure*.

excess risk. The difference between the incidence of a specified *stochastic effect* observed in an exposed group to that in an unexposed *control* group.

lifetime risk. The probability that a specified *health effect* will occur at some time in the future in an individual as a result of *radiation exposure*.

relative risk. The ratio between the incidence of a specified *stochastic effect* observed in an exposed group and that in an unexposed *control* group. (See *control* (2).)

4.   ***radiation risks.*** See *radiation risks*.

5.   The potential for an unwanted outcome resulting from a *nuclear security event* as determined by its likelihood and the associated consequences.

> ! A number of IAEA Nuclear Security Series publications, notably Ref. [6], refer to different 'types' of *risk* relevant to *nuclear security*, specifically: the *risk* of unauthorized removal of radioactive material (with possible subsequent dispersal or use to cause radiation exposure or, in the case of nuclear material, with the intent to construct a nuclear explosive device); and the *risk* of sabotage. This wording could be understood to be used in a loose and general way, but possibly more specifically to indicate that successful unauthorized removal of material or sabotage would itself be a 'consequence', in which case the *risk* would be rather vaguely that to society or the world in general. The descriptions in the text, however, indicate that the 'types' are actually the *risk* associated with unauthorized removal and the *risk* associated with sabotage, i.e. the risks to persons, society, property or the environment that might be affected by a malicious act using removed material or releases from a sabotaged facility.

> § In *safety*, quantitative probabilities (based on observed or modelled frequencies of random events) are often used to calculate *risk*: in *security* contexts, likelihoods are very unlikely to be quantifiable, as they usually depend on human decisions and actions rather than random processes.

risk assessment

See *assessment* (1).

risk coefficient, γ

The *lifetime risk* or *radiation detriment* assumed to result from *exposure* to unit *equivalent dose* or *effective dose*.

risk constraint

A prospective and *source* related value of individual *risk* that is used in *planned exposure situations* as a parameter for the *optimization of protection and safety* for the *source*, and that serves as a boundary in defining the range of options in *optimization*.

⓪ The *risk constraint* is a *source* related value that provides a basic level of *protection* for the individuals most at *risk* from a *source*.

⓪ This *risk* is a function of the probability of an unintended event causing a *dose*, and the probability of the detriment due to such a *dose*.

⓪ *Risk constraints* correspond to *dose constraints* but apply to *potential exposure*.

[risk factor]

! Sometimes misused as a synonym for *risk coefficient*. This is different from the normal medical use of the term *risk factor* to indicate a factor that influences an individual's *risk*, and its use as a synonym for *risk coefficient* should be avoided.

! *Risk factor* should be used only in the medical sense.

risk monitor

A plant specific real time *analysis* tool used to determine the instantaneous *risk* based on the actual status of the *systems* and *components*.

⓪ At any given time, the *risk monitor* reflects the current plant configuration in terms of the known status of the various *systems* and/or *components*; for example, whether there are any *components* out of service for *maintenance* or tests.

⓪ The *model* used by the *risk monitor* is based on, and is consistent with, the *'living' probabilistic safety assessment* for the *facility*.

risk projection model

See *model*.

rock, igneous

See *igneous rock*.

[röntgen (R)]

Unit of *exposure*, equal to 2.58×10^{-4} C/kg (exactly).

⓪ Superseded by the SI unit C/kg.

root cause

See *cause*.

root uptake

See *uptake* (1).

routine monitoring

See *monitoring* (1).

runup

A sudden surge of water up a beach or a *structure*.

S

sabotage

1.   Any deliberate act directed against a *nuclear facility* or *nuclear material* in use, storage or transport which could directly or indirectly endanger the health and safety of personnel, the public or the environment by exposure to radiation or release of radioactive substances.

> ! This definition of sabotage is of a technical nature and does not aim to provide a definition for the purpose of criminal law, such as those provided for in the relevant international instruments or national law of States.

> ⓘ This definition is the same as that in the 2005 CPPNM Amendment [5].

2.   Any deliberate act directed against an *associated facility* or an *associated activity* that could directly or indirectly endanger the health and safety of personnel, the public, or the environment by exposure to radiation or release of *radioactive substances*.

safe state

See *plant states (considered in design)*.

safeguards agreement

An agreement for the application of safeguards concluded between the IAEA and a State or a group of States, in some cases together with a regional authority responsible for safeguards implementation, such as the European Atomic Energy Community (Euratom) and the Brazilian–Argentine Agency for Accounting and Control of Nuclear Materials (ABACC). Such an agreement is concluded either because of the requirements of a project and supply agreement, or to satisfy the relevant requirements of bilateral or multilateral arrangements, or to apply safeguards at the request of a State to any of that State's nuclear activities.

> ⓘ See the Safeguards Glossary [14].

safety

See *(nuclear) safety* and *protection and safety*.

> ⓘ In the Fundamental Safety Principles (IAEA Safety Fundamentals), the generalized usage in this particular text of the term *safety* (i.e. to mean *protection and safety*) is explained as follows (SF-1 [24], paras 3.1 and 3.2):

> "3.1. For the purposes of this publication, '*safety*' means the *protection* of people and the *environment* against *radiation risks*, and the *safety* of *facilities and activities* that give rise to *radiation risks*. '*Safety*' as used here and in the IAEA *safety standards* includes the *safety* of *nuclear installations*, *radiation safety*, the *safety* of *radioactive waste management* and *safety* in the *transport* of *radioactive material*; it does not include non-radiation-related aspects of *safety*.

> "3.2. *Safety* is concerned with both *radiation risks* under normal circumstances and *radiation risks* as a consequence of *incidents*[4], as well as with other possible direct consequences of a loss of control over a nuclear reactor core, nuclear chain reaction, *radioactive source* or any other *source* of radiation. *Safety measures* include actions to prevent *incidents* and arrangements put in place to mitigate their consequences if they were to occur.

> " [4] '*Incidents*' includes *initiating events*, *accident precursors*, *near misses*, *accidents* and unauthorized acts (including malicious *acts* and non-malicious acts)."

safety action

A single action taken by a *safety actuation system*.

ⓘ For example, insertion of a *control* rod, closing of *containment* valves or *operation* of the *safety* injection pumps.

safety actuation system

See *plant equipment (for a nuclear power plant)*.

safety analysis

See *analysis*.

safety assessment

See *assessment* (1).

safety case

A collection of arguments and evidence in support of the *safety* of a *facility or activity*.

ⓘ This will normally include the findings of a *safety assessment* and a statement of confidence in these findings.

ⓘ For a *disposal facility*, the *safety case* may relate to a given stage of development. In such cases, the *safety case* should acknowledge the existence of any unresolved issues and should provide guidance for work to resolve these issues in future development stages.

safety categorization

For nuclear power plants, the categorization into a limited number of safety categories of the functions that are required for fulfilling the main *safety functions* in different *plant states*, including all modes of *normal operation*, on the basis of their safety significance.

ⓘ See Refs [25, 54].

safety class

See *safety classification*.

safety classification

For nuclear power plants, the assignment to a limited number of *safety classes* of *systems* and *components* and other items of equipment on the basis of their functions and their *safety* significance.

safety class. For nuclear power plants, the classes into which *systems* and *components* and other items of equipment are assigned on the basis of their functions and their *safety* significance.

ⓘ The *design* is required to ensure in particular that any *failure* of *items important to safety* in a *system* in a lower *safety class* will not propagate to a *system* in a higher *safety class*. Items of equipment that perform multiple functions are required to be classified in a *safety class* that is consistent with the most important function performed by the items of equipment.

ⓘ See Requirement 22 of SSR-2/1 (Rev. 1) [25] and para. 2.2 of SSG-30 [54].

safety committee

A group of experts convened by the *operating organization* to advise on the *safety* of *operation* of an *authorized facility*.

safety culture

The assembly of characteristics and attitudes in organizations and individuals which establishes that, as an overriding priority, *protection and safety issues* receive the attention warranted by their significance.

> ⓘ For a more detailed discussion, see Ref. [55].

See also *nuclear security culture*.

safety feature for design extension conditions

See *plant equipment (for a nuclear power plant)*.

safety function

A specific purpose that must be accomplished for *safety* for a *facility* or *activity* to prevent or to mitigate radiological consequences of *normal operation*, *anticipated operational occurrences* and *accident conditions*. (See SSG-30 [54].)

> ⓘ SSR-2/1 (Rev. 1) [25] establishes *requirements* on *safety functions* to be fulfilled by the *design* of a nuclear power plant in order to meet three general *safety requirements*:
>
> (a) The capability to safely shut down the reactor and maintain it in a safe *shutdown* condition during and after appropriate *operational states* and *accident conditions*;
>
> (b) The capability to remove *residual heat* from the reactor core, the reactor and *nuclear fuel* in storage after *shutdown*, and during and after appropriate *operational states* and *accident conditions*;
>
> (c) The capability to reduce the potential for the *release* of *radioactive material* and to ensure that any *releases* are within *prescribed limits* during and after *operational states* and within *acceptable limits* during and after *design basis accidents*.

This guidance is commonly condensed into a succinct expression of three ***fundamental safety functions*** for nuclear power plants:

(a) *Control* of *reactivity*;

(b) Cooling of *radioactive material*;

(c) *Confinement* of *radioactive material*.

In some *IAEA publications*, 'basic *safety function*' and '***main safety function***' are also used.

safety group

The assembly of equipment designated to perform all actions required for a particular *initiating event* to ensure that the *limits* specified in the *design basis* for *anticipated operational occurrences* and *design basis accidents* are not exceeded.

! The term 'group' is also used (with various qualifying adjectives, e.g. *maintenance* group, *commissioning* group) in the more obvious sense of a group of people involved in a particular area of work. Such terms may need to be defined if there is any chance of confusion with *safety group*.

safety indicator

A quantity used in *assessments* as a measure of the radiological impact of a *source* or of a *facility or activity*, or of the performance of *protection and safety* provisions, other than a prediction of *dose* or *risk*.

ⓘ Such quantities are most commonly used in situations where predictions of *dose* or *risk* are unlikely to be reliable; for example, long term *assessments* of *repositories*.

ⓘ They are normally either:

(a) Illustrative calculations of *dose* or *risk* quantities, used to give an indication of the possible magnitude of *doses* or *risks* for comparison with criteria; or

(b) Other quantities, such as radionuclide concentrations or fluxes, that are considered to give a more reliable indication of impact, and that can be compared with other relevant data.

safety issues

Deviations from current *safety standards* or *practices*, or weaknesses in *facility design* or *practices* as identified by plant *events*, with a potential impact on *safety* because of their impact on *defence in depth, safety* margins or *safety culture*.

safety layers

Passive *systems*, automatically or manually initiated *safety systems*, or administrative *controls* that are provided to ensure that the required *safety functions* are achieved.

ⓘ Often expressed as:

(a) Hardware (i.e. passive and active *safety systems*);

(b) Software, including personnel and *procedures* as well as computer software;

(c) Management *control*, in particular preventing *defence in depth* degradation (through quality management, *preventive maintenance, surveillance testing*, etc.) and reacting appropriately to experience feedback from degradations that do occur (e.g. determining *root causes*, taking corrective actions).

See also *defence in depth (1)*.

safety limits

See *limit*.

safety measure

Any action that might be taken, condition that might be applied or *procedure* that might be followed to fulfil the *requirements* of Safety Requirements.

safety (of radioactive sources)

[Measures intended to minimize the likelihood of *accidents* involving *radioactive sources* and, should such an *accident* occur, to mitigate its consequences.] (See Ref. [21].)

   ! This definition is particular to the Code of Conduct on the Safety and Security of Radioactive Sources [21].

safety related item

See *plant equipment (for a nuclear power plant)*.

safety related system

See *plant equipment (for a nuclear power plant)*.

safety standards

Standards issued pursuant to Article III(A)(6)[12] of the Statute of the IAEA [47].

   (i) *Requirements*, regulations, standards, rules, codes of practice or recommendations established to protect people and the *environment* against *ionizing radiation* and to minimize danger to life and property.

   (i) *Safety standards* issued since 1997 in the IAEA Safety Standards Series are designated as Safety Fundamentals, Safety Requirements or Safety Guides.

   (i) Some *safety standards* issued prior to 1997 in the (defunct) Safety Series were designated Safety Standards, Codes, Regulations or Rules.

   (i) Furthermore, some publications issued in the (defunct) Safety Series were not *safety standards*, notably those designated Safety Practices or Procedures and Data.

   (i) Other *IAEA publications*, such as Safety Reports and TECDOCs (most of which are issued pursuant to Article VIII of the Statute), are not *safety standards*.

safety system

See *plant equipment (for a nuclear power plant)*.

safety system settings

See *plant equipment (for a nuclear power plant)*.

safety system support features

See *plant equipment (for a nuclear power plant)*.

---

[12] "[The Agency is authorized] To establish or adopt, in consultation and, where appropriate, in collaboration with the competent organs of the United Nations and with the specialized agencies concerned, standards of safety for protection of health and minimization of danger to life and property (including such standards for labour conditions)…"

safety task

The sensing of one or more variables indicative of a specific *postulated initiating event*, the signal processing, the initiation and completion of the *safety actions* required to prevent the *limits* specified in the *design basis* from being exceeded, and the initiation and completion of certain services of the *safety system support features*.

scenario

A postulated or assumed set of conditions and/or *events*.

ⓘ Most commonly used in *analysis* or *assessment* to represent possible future conditions and/or *events* to be modelled, such as possible *accidents* at a *nuclear facility*, or the possible future evolution of a *disposal facility* and its surroundings. A *scenario* may represent the conditions at a single point in time or a single *event*, or a time history of conditions and/or *events* (including *processes*).

ⓘ See *event*.

attack scenario. A postulated or assumed set of conditions and events, commonly used in analysis or assessment to represent possible future conditions and events to be modelled, such as a possible nuclear security event.

reference scenario. A hypothetical but possible evolution of a *disposal facility* and its surroundings on the basis of activities, such as construction work, mining or drilling, that have a high probability of being undertaken by people in the future and that could cause a *human intrusion* into the *disposal facility*, and which can be evaluated.

scram

A rapid *shutdown* of a nuclear reactor in an *emergency*.

See also *anticipated transient without scram (ATWS)*.

screening

A type of *analysis* aimed at eliminating from further consideration factors that are less significant for *protection* or *safety* in order to concentrate on the more significant factors.

ⓘ This is typically achieved by consideration of very pessimistic hypothetical *scenarios*.

ⓘ *Screening* is usually conducted at an early stage in order to narrow the range of factors needing detailed consideration in an *analysis* or *assessment*.

screening distance value (SDV)

The distance from a *facility* beyond which, for *screening* purposes, potential origins of a particular type of *external event* can be ignored.

screening probability level (SPL)

A value of the annual probability of occurrence of a particular type of *event* below which, for *screening* purposes, such an *event* can be ignored.

seabed disposal

See *disposal* (3).

sealed source

See *source* (2).

[secondary limit]

See *limit*.

secondary waste

See *waste*.

security

See *nuclear security* (1).

segregation

1.   See *radioactive waste management* (1).

2.   The physical separation of *structures, systems and components* by distance or by means of some form of *barrier* to reduce the likelihood of *common cause failures*.

3.   Separation of *transport packages* from persons, undeveloped photographic film and dangerous goods and separation of *transport packages* containing *fissile material* from each other. (See SSR-6 (Rev. 1) [2].)

seismic demand

The applicable seismic load for a *structure, system or component*.

   ⓘ   Typically, the seismic demand is expressed in terms of acceleration time history, acceleration *response spectra*, and seismic induced forces and/or displacements.

seismic qualification

See *qualification*: *equipment qualification*.

seismogenic structure

A structure that displays earthquake activity or that manifests historical surface rupture or the effects of *palaeoseismicity*, and is likely to generate macro-earthquakes within a time period of concern.

seismotectonic model

See *model*.

self-assessment

See *assessment* (2).

senior management

The person or persons who direct, control and assess an organization at the highest level.

sensitive digital assets

See *sensitive information*.

sensitive information

Information, in whatever form, including software, the unauthorized disclosure, modification, alteration, destruction, or denial of use of which could compromise nuclear security.

sensitive digital assets. *Sensitive information assets* that are (or are parts of) *computer-based systems*.

ⓘ Alternatively, these are *digital assets* that store, process, control or transmit *sensitive information*.

See also *digital assets*.

sensitive information assets. *Any equipment or components that are used to store, process, control or transmit sensitive information.*

ⓘ For example, *sensitive information assets* include control systems, networks, information systems and any other electronic or physical media. sensitive information assets.

See *sensitive information*.

sensitivity analysis

See *analysis*.

service conditions

Physical conditions prevailing or expected to prevail during the *service life* of a *structure, system or component*.

ⓘ *Service conditions* include environmental conditions (e.g. conditions of humidity; thermal, chemical, electrical, mechanical and radiological conditions), and *operating conditions* (conditions in *normal operation*, error induced conditions) and conditions during and after events.

specified service conditions. Physical conditions and stressors to which equipment is subjected to during its service life.

ⓘ This includes normal operating conditions, process conditions, abnormal operating conditions and conditions during and following a *design basis accident* or in *design extension conditions*.

service life

See *life, lifetime*.

severe accident

See *accident* (1).

severe accident management

See *accident management*.

severe deterministic effect

See *health effects (of radiation)*: *deterministic effect*.

sheltering

The short term use of a structure for *protection* from an airborne plume and/or deposited *radioactive* material.

    ⓘ Sheltering is an *urgent protective action*, used to provide shielding against *external exposure* and to reduce the *intake* of airborne radionuclides by inhalation.

shipment

The specific movement of a *consignment* from origin to destination. (See SSR-6 (Rev. 1) [2].)

shipper

See *consignor*.

short lived waste

See *waste classes*.

shutdown

The cessation of *operation* of a *facility*.

    permanent shutdown. The cessation of *operation* of a *facility* with no intention to recommence *operation* in the future.

    ⓘ Between the *permanent shutdown* of the *facility* and approval of the *decommissioning plan*, there may be a period of transition.

    ⓘ During such a transition period, the *authorization* for *operation* of the *facility* remains in place unless the *regulatory body* has approved modifications to the *authorization* on the basis of a reduction in the *hazards* associated with the *facility*.

    ⓘ During this transition period, some preparatory actions for *decommissioning* can be performed in accordance with the *authorization* for *operation* of the *facility* or a modified *authorization*.

shutdown reactivity

See *reactivity*.

sievert (Sv)

The SI unit of *equivalent dose* and *effective dose*, equal to 1 J/kg.

signature

A characteristic or a set of characteristics of a given sample that enables that sample to be compared with reference materials.

significant transboundary release

A *release* of *radioactive material* to the *environment* that may result in *doses* or levels of *contamination* beyond national borders from the *release* which exceed *generic criteria* for *protective actions* and *other response actions*, including *food* restrictions and restrictions on trade.

single failure

A *failure* which results in the loss of capability of a single *system* or *component* to perform its intended *safety function(s)*, and any consequential *failure(s)* which result from it.

single failure criterion

A criterion (or *requirement*) applied to a *system* such that it must be capable of performing its task in the presence of any *single failure*.

> ⓘ To ensure that the *single failure criterion* is met, usually two or more independent (*redundant*) *systems* or trains are provided by design to achieve the same *safety function*.

> ***double contingency principle.*** A principle applied, for example, in the *design* of *processes* for *nuclear fuel cycle facilities*, such that the *design* for a *process* must include sufficient safety features that a *criticality accident* would not be possible unless at least two unlikely and independent changes in *process* conditions were to occur concurrently.

site area

See *area*.

site area emergency

See *emergency class*.

site boundary

See *area: site area*.

site characterization

See *characterization* (2).

site confirmation (in the siting process for a disposal facility)

The final stage of the *siting process* for a *disposal facility*, based on detailed investigations on the preferred site which provide site specific information needed for *safety assessment*.

ⓘ This stage includes the finalization of the *design* for the *disposal facility* and the preparation and submission of a *licence* application to the *regulatory body*.

ⓘ *Site confirmation* follows *site characterization* for a disposal facility.

site evaluation

Analysis of those factors at a site that could affect the *safety* of a *facility or activity* on that site.

ⓘ This includes *site characterization*, consideration of factors that could affect safety features of the *facility* or *activity* so as to result in a *release* of *radioactive material* and/or could affect the *dispersion* of such material in the *environment*, as well as population and access issues relevant to *safety* (e.g. feasibility of *evacuation*, location of people and resources).

ⓘ The *analysis* for a site of the origins of *external events* that could give rise to *hazards* with potential consequences for the *safety* of a nuclear power plant constructed on that site.

ⓘ For a nuclear power plant, *site evaluation* typically involves the following stages:

(a) *Site selection* stage. One or more preferred candidate sites are selected after the investigation of a large region, the rejection of unsuitable sites, and *screening* and comparison of the remaining sites.

(b) *Site characterization* stage. This stage is further subdivided into:

—**Site verification**, in which the suitability of the site to host a nuclear power plant is verified mainly according to predefined site *exclusion* criteria;

—Site confirmation, in which the characteristics of the site necessary for the purposes of *analysis* and detailed *design* are determined.

(c) Pre-operational stage. Studies and investigations begun in the previous stages are continued after the start of *construction* and before the start of *operation* of the plant, to complete and refine the *assessment* of site characteristics. The site data obtained allow a final *assessment* of the simulation *models* used in the final *design*.

(d) Operational stage. Appropriate *safety* related *site evaluation activities* are carried out throughout the *lifetime* of the *facility*, mainly by means of *monitoring* and *periodic safety review*.

site personnel

All persons working in the *site area* of an *authorized facility*, either permanently or temporarily.

site (seismic) response

The behaviour of a rock column or soil column at a site under a prescribed ground motion load.

site selection

See *siting*.

site survey

See *siting*.

site verification

See *site evaluation*.

siting

The *process* of selecting a suitable site for a *facility*, including appropriate *assessment* and definition of the related *design bases*.

(i) The *siting process* for a *nuclear installation* generally consists of *site survey* and *site selection*.

site survey. The *process* of identifying candidate sites for a *nuclear installation* after the investigation of a large region and the rejection of unsuitable sites.

site selection. The *process* of assessing the remaining sites by *screening* and comparing them on the basis of *safety* and other considerations to select one or more preferred candidate sites.

See also *site evaluation*.

(i) The *siting process* for a *disposal facility* is particularly crucial to its long term *safety*; it may therefore be a particularly extensive *process*, and is divided into the following stages:

— Concept and planning;

— *Area survey*;

— *Site characterization*;

— *Site confirmation*.

(i) The terms *siting*, *design*, *construction*, *commissioning*, *operation* and *decommissioning* are normally used to delineate the six major stages of the *lifetime* of an *authorized facility* and of the associated *licensing process*. In the special case of *disposal facilities* for *radioactive waste*, *decommissioning* is replaced in this sequence by *closure*.

situation awareness

The dynamic process of perception and comprehension of the plant's actual condition in order to support the ability of individuals and teams to predict the future conditions of systems.

(i) *Situation awareness* is a way of forming a mental model of the situation and future planned actions. The degree of *situation awareness* corresponds to the difference between the understanding of plant conditions and the actual conditions at any given time. One of the objectives of *human factors engineering* is to support the development of *situation awareness* in *operating personnel*.

SL-1, SL-2

Levels of ground motion (representing the potential effects of earthquakes) considered in the *design basis* for a *facility*.

(i) *SL-1* corresponds to a less severe, more likely earthquake than *SL-2*.

(i) In some States, *SL-1* corresponds to a level with a probability of 10^{-2} per year of being exceeded, and *SL-2* corresponds to a level with a probability of 10^{-4} per year of being exceeded.

small freight container

See *freight container*.

somatic effect

See *health effects (of radiation)*.

sorption

The interaction of an atom, molecule or particle with the solid surface at a solid–solution or a solid–gas interface.

ⓘ Used in the context of radionuclide *migration* to describe the interaction of radionuclides in pore water or groundwater with soil or host rock, and of radionuclides in surface water bodies with suspended and bed sediments.

ⓘ A general term which includes **absorption** (interactions taking place largely within the pores of solids) and **adsorption** (interactions taking place on solid surfaces).

ⓘ The *processes* involved can also be divided into **chemisorption** (chemical bonding with the substrate) and **physisorption** (physical attraction, e.g. by weak electrostatic forces).

ⓘ In practice, *sorption* may sometimes be difficult to distinguish from other factors affecting *migration*, such as filtration or *dispersion*.

source

1.   Anything that may cause *radiation exposure* — such as by emitting *ionizing radiation* or by releasing *radioactive substances* or *radioactive material* — and can be treated as a single entity for purposes of *protection and safety*.

ⓘ For example, materials emitting *radon* are *sources* in the *environment*; a sterilization gamma irradiation unit is a *source* for the *practice* of irradiation preservation of *food* and sterilization of other products; an X ray unit may be a *source* for the *practice* of radiodiagnosis; a nuclear power plant is part of the *practice* of generating electricity by nuclear fission, and may be regarded as a *source* (e.g. with respect to *discharges* to the *environment*) or as a collection of *sources* (e.g. for occupational *radiation protection* purposes).

ⓘ A complex or multiple installation situated at one location or site may, as appropriate, be considered a single *source* for the purposes of application of *safety standards*.

natural source. A naturally occurring *source* of *radiation*, such as the sun and stars (*sources* of cosmic *radiation*) and rocks and soil (terrestrial *sources* of *radiation*), or any other material whose *radioactivity* is for all intents and purposes due only to *radionuclides of natural origin*, such as products or residues from the processing of minerals; but excluding *radioactive material* for use in a *nuclear installation* and *radioactive waste* generated in a *nuclear installation*.

ⓘ Examples of *natural sources* include *naturally occurring radioactive material (NORM)* associated with the processing of raw materials (e.g. feedstocks, intermediate products, final products, co-products, *waste*).

radiation generator. A device capable of generating *ionizing radiation*, such as X rays, neutrons, electrons or other charged particles, that may be used for scientific, industrial or medical purposes.

See also *radiation exposure device*.

radiation source. [A *radiation generator*, or a *radioactive source* or other *radioactive material* outside the *nuclear fuel cycles* of research and power reactors.]

! Defined in the 2001 edition of the Code of Conduct on the Safety and Security of Radioactive Sources, but not included in the 2004 edition (see Ref. [21]).

2. *Radioactive material* used as a *source* of *radiation*.

ⓘ Such as those *sources* used for medical applications or in industrial instruments. These are, of course, *sources* as defined in (1), but this usage in (2) is less general.

dangerous source. A *source* that could, if not under *control*, give rise to *exposure* sufficient to cause *severe deterministic effects*. This categorization is used for determining the need for *emergency arrangements* and is not to be confused with categorizations of *sources* for other purposes.

ⓘ The term *dangerous source* relates to dangerous quantities of *radioactive* material (D values) as recommended in Ref. [56].

disused source. A radioactive source that is no longer used, and is not intended to be used, for the practice for which an authorization has been granted. (See Ref. [21].)

! Note that a *disused source* may still represent a significant radiological *hazard*. It differs from a *spent source* in that it may still be capable of performing its function: it may be disused because it is no longer needed.

ⓘ The Joint Convention on the Safety of Spent Fuel Management and on the Safety of Radioactive Waste Management [11] refers to "disused sealed sources", but does not define them.

disused sealed source. A *radioactive source*, comprising *radioactive material* that is permanently sealed in a capsule or closely bonded and in a solid form (excluding reactor *fuel elements*), that is no longer used, and is not intended to be used, for the *practice* for which an *authorization* was granted.

ⓘ The definition is provided on the basis of the definition of *disused source* (see above) and the definition of *sealed source* (see below).

orphan source. A *radioactive source* which is not under *regulatory control*, either because it has never been under *regulatory control* or because it has been abandoned, lost, misplaced, stolen or otherwise transferred without proper *authorization*. (See Ref. [21].)

radioactive source

1. A *source* containing *radioactive material* that is used as a *source* of *radiation*.

2. *Radioactive material* that is permanently sealed in a capsule or closely bonded and in a solid form and which is not exempt from regulatory control. This also includes any *radioactive material* released if the *radioactive source* is leaking or broken, but does not include material encapsulated for *disposal*, or *nuclear material* within the *nuclear fuel cycles* of research and power reactors.

ⓘ   This definition is from the Code of Conduct on the Safety and Security of Radioactive Sources [21].

sealed source. A *radioactive source* in which the *radioactive material* is (a) permanently sealed in a capsule or (b) closely bonded and in a solid form.

ⓘ The Joint Convention on the Safety of Spent Fuel Management and on the Safety of Radioactive Waste Management definition [11] is *radioactive material* that is (a) permanently sealed in a capsule or (b) closely bonded and in a solid form, excluding reactor *fuel elements*.

ⓘ The term *special form radioactive material*, used in the context of *transport* of *radioactive material*, has essentially the same meaning.

ⓘ Definition 2 of *radioactive source*, which is taken from the Code of Conduct on the Safety and Security of Radioactive Sources [21], has essentially the same meaning.

ⓘ *Disused sealed source*: see *source*: *disused source*.

spent source. A *source* that is no longer suitable for its intended purpose as a result of *radioactive* decay.

! Note that a *spent source* may still represent a radiological *hazard*.

unsealed source. A *radioactive source* in which the *radioactive material* is neither (a) permanently sealed in a capsule nor (b) closely bonded and in a solid form.

vulnerable source. A *radioactive source* for which the *control* is inadequate to provide assurance of long term *safety* and *security*, such that it could relatively easily be acquired by unauthorized persons.

source material

See *nuclear material* (2).

source monitoring

See *monitoring* (1).

source region

A region within the body containing one or more radionuclides.

ⓘ Used in internal dosimetry; for example, for radionuclides irradiating a *target tissue*.

source term

The amount and isotopic composition of *radioactive material* released (or postulated to be released) from a *facility*.

ⓘ Used in modelling *releases* of radionuclides to the *environment*, in particular in the context of *accidents* at *nuclear installations* or *releases* from *radioactive waste* in *repositories*.

special arrangement

Those provisions, approved by the *competent authority*, under which *consignments* that do not satisfy all the applicable *requirements* of [the Transport] Regulations may be transported. (See SSR-6 (Rev. 1) [2].)

special facility

A facility for which predetermined facility specific actions need to be taken if *urgent protective actions* are ordered in its locality in the event of a *nuclear or radiological emergency*.

ⓘ Examples include chemical plants that cannot be evacuated until certain actions have been taken to prevent fire or explosions and telecommunications centres that must be staffed in order to maintain telephone services.

ⓘ This is not necessarily a *facility* within the meaning of the term *facilities and activities*.

special fissionable material

See *nuclear material* (2).

special form radioactive material

Either an indispersible solid *radioactive material* or a sealed capsule containing *radioactive material*. (See SSR-6 (Rev. 1) [2].)

special monitoring

See *monitoring* (1).

special population group

Members of the public for whom special arrangements are necessary in order for effective *protective actions* to be taken in the event of a *nuclear or radiological emergency*. Examples include disabled persons, hospital *patients* and prisoners.

specific activity

See *activity* (1): *specific activity*.

spent fuel

1.   *Nuclear fuel* removed from a reactor following irradiation that is no longer usable in its present form because of depletion of *fissile material*, *poison* buildup or *radiation* damage.

> ⓘ The participle 'spent' suggests that *spent fuel* cannot be used as *fuel* in its present form (e.g. as in *spent source*). In practice, however (as in (2) below), *spent fuel* is commonly used to refer to *fuel* that has been used as *fuel* but will no longer be used, whether or not it could be used (and that might more accurately be termed 'disused *fuel*').

2.   [*Nuclear fuel* that has been irradiated in and permanently removed from a reactor core.] (See Ref. [11].)

spent fuel management

All *activities* that relate to the handling or storage of *spent fuel*, excluding *off-site* transport. It may also involve *discharges*. (See Ref. [11].)

spent fuel management facility

Any *facility* or installation the primary purpose of which is *spent fuel management*. (See Ref. [11].)

spent source

See *source* (2).

[stakeholder]

See *interested party*.

> ! The term *stakeholder* is used in the same broad sense as *interested party* and the same provisos are necessary. The term *stakeholder* has disputed usages and is misleading and too all-encompassing for clear use. In view of the potential for misunderstanding and misrepresentation, use of the term is discouraged in favour of *interested party*.

> ⓘ To 'have a stake in' something, figuratively, means to have something to gain or lose by, or to have an interest in, the turn of events.

> ⓘ The Handbook on Nuclear Law [43] states that: "Owing to the differing views on who has a genuine interest in a particular nuclear related activity, no authoritative definition of stakeholder has yet been offered, and no definition is likely to be accepted by all parties."

standards dosimetry laboratory

A laboratory, designated by the relevant national authority, that possesses *certification* or accreditation necessary for the purpose of developing, maintaining or improving primary or secondary standards for radiation dosimetry.

stand-off attack

An attack, executed at a distance from the *target* nuclear facility or transport, which does not require adversary hands-on access to the *target*, or require the adversary to overcome the physical protection system.

station blackout

Plant condition with complete loss of all AC power from off-site sources, from the main generator and from standby AC power sources important to safety to the essential and non-essential switchgear buses.

> ⓘ DC power supplies and uninterruptible AC power supplies may be available as long as batteries can supply the loads. *Alternate AC power supplies* are available.

static analysis

See analysis.

State of destination

A State to which a *transboundary movement* is planned or takes place. (See Ref. [11].)

State of origin

A State from which a *transboundary movement* is planned to be initiated or is initiated. (See Ref. [11].)

State of transit

Any State, other than a *State of origin* or a *State of destination*, through whose territory a *transboundary movement* is planned or takes place. (See Ref. [11].)

stochastic analysis

See *probabilistic analysis*.

stochastic effect

See *health effects (of radiation)*.

storage

The holding of *radioactive sources*, *radioactive material*, *spent fuel* or *radioactive waste* in a *facility* that provides for their/its *containment*, with the intention of retrieval.

    ⓘ Generalized from the Joint Convention on the Safety of Spent Fuel Management and on the Safety of Radioactive Waste Management [11], the Code of Conduct on the Safety and Security of Radioactive Sources [21] and GSR Part 5 [57].

    ! *Storage* is by definition an interim measure, and the term **[interim storage]** would therefore be appropriate only to refer to short term temporary *storage* when contrasting this with the longer term fate of the *waste*.

    ! *Storage* as defined above should not be described as *interim storage*.

    ! In many cases, the only element of this definition that is important is the distinction between *disposal* (with no intent to retrieve) and *storage* (with intent to retrieve).

    ⓘ In such cases, a definition is not necessary; the distinction can be made in the form of a footnote at the first use of the term *disposal* or *storage* (e.g. "Use of the term *disposal* indicates that there is no intention to retrieve the *waste*. If retrieval of the *waste* at any time in the future is intended, the term *storage* is used.").

    ⓘ For *storage* in a combined *storage* and *disposal facility*, for which a decision may be made at the time of its *closure* whether to remove the *waste* stored during the *operation* of the *storage facility* or to dispose of it by encasing it in concrete, the question of intention of retrieval may be left open until the time of *closure* of the *facility*.

    ⓘ Contrasted with *disposal*.

    dry storage. *Storage* in a gaseous medium, such as air or an inert gas.

    ⓘ *Dry storage facilities* include *facilities* for the *storage* of *spent fuel* in casks, silos or vaults.

    wet storage. *Storage* in water or in another liquid.

    ⓘ The universal mode of *wet storage* consists of storing *spent fuel* assemblies or *spent fuel* elements in pools of water or other liquids, usually supported on racks or in baskets and/or in *canisters* that also contain liquid.

    ⓘ The liquid in the pool surrounding the *fuel* provides for heat dissipation and *radiation* shielding, and the racks or other devices ensure a geometrical configuration that maintains subcriticality.

strategic location

A location of high security interest in the State which is a potential *target* for terrorist attacks using *nuclear material* or *other radioactive material*, or a location at which *nuclear material* or *other radioactive material* that is *out of regulatory control* is located.

ⓘ In some publications in the IAEA Nuclear Security Series, this has also been defined as "A location of high security interest in the State which is a potential *target* for terrorist attacks using *nuclear and other radioactive material* or a location for *detection* of *nuclear and other radioactive material* that is out of *regulatory control*" [8].

strombolian eruption

See volcanic *eruption*.

strongly penetrating radiation

See *radiation*.

structure

See *structures, systems and components*.

structures, systems and components (SSCs)

A general term encompassing all of the elements (items) of a *facility* or *activity* that contribute to *protection and safety*, except human factors.

ⓘ Human factors may be reflected in *structures, systems and components* in so far as ergonomics — the study of people's efficiency in their work setting — is an element in their design.

See also *core components*.

component. One of the parts that make up a *system*.

ⓘ A *component* may be a hardware *component* (e.g. wires, transistors, integrated circuits, motors, relays, solenoids, pipes, fittings, pumps, tanks, valves) or a software *component* (e.g. modules, routines, programmes, software functions).

ⓘ A *component* may be made up of other *components*.

ⓘ The terms 'equipment', '*component*' and 'module' are often used interchangeably. The relationship of these terms is not yet standardized.

See also *active component*, *passive component* and *core components*.

structure. A passive element (e.g. buildings, vessels, shielding).

system. A set of *components* which interact according to a *design* so as to perform a specific (active) function, in which an element of the *system* can be another *system*, called a subsystem.

ⓘ Examples are mechanical *systems*, electrical *systems* and instrumentation and *control systems*.

sub-seabed disposal

See *disposal* (1).

substantiation

See *qualification*: *equipment qualification*.

supervised area

See *area*.

supplier (of a source)

Any *person or organization* to whom a *registrant* or *licensee* assigns duties, totally or partially, in relation to the *design*, manufacture, production or *construction* of a *source*.

ⓘ An importer of a *source* is considered a *supplier* of the *source*.

ⓘ The term *supplier* (of a *source*) includes designers, manufacturers, producers, constructors, assemblers, installers, distributors, sellers, importers or exporters of a *source*.

surface contaminated object (SCO)

A solid object that is not itself *radioactive* but which has *radioactive material* distributed on its surface. (See SSR-6 (Rev. 1) [2].)

! This usage is specific to the Transport Regulations [2], and should otherwise be avoided.

surface faulting

Permanent offsetting or tearing of the ground surface by differential movement across a fault in an earthquake.

surveillance

1. A type of *inspection* to verify the integrity of a *facility* or structure.

ⓘ For example, *surveillance* is used in the context of a *disposal facility* for *radioactive waste* to mean physical inspection of the *facility* to verify its integrity and the capability to protect and preserve passive *barriers*.

2. The collection of information through devices or direct observation to detect unauthorized movements of *nuclear material*, tampering with *containment* of *nuclear material* or falsification of information related to location and quantities of *nuclear material*.

ⓘ This definition is for use in relation to material *out of regulatory control*.

surveillance testing

Periodic testing to verify that *structures, systems and components* continue to function or are capable of performing their functions when called upon to do so.

survey

> ***area survey.*** An early stage of the *siting process* for a *disposal facility*, during which a broad region is examined to eliminate unsuitable areas and to identify other areas which may contain suitable sites.
>
> ⓘ *Area survey* is followed by *site characterization*.
>
> ⓘ *Area survey* may also refer to the *siting process* for any other *authorized facility*.

See also *site evaluation*, which includes *site characterization* and is not specific to a *disposal facility* site.

habit survey. An evaluation of those aspects of the behaviour of *members of the public* that might influence their *exposure* — such as diet, *food* consumption rates or occupancy of different areas — usually aimed at characterizing the *representative person*.

system

See *structures, systems and components*.

system code

A *computational model* that is capable of simulating the transient performance of a complex *system* such as a nuclear power plant.

> ⓘ A *system code* typically includes equations for thermohydraulics, neutronics and heat transfer, and must include special *models* for simulating the performance of *components* such as pumps and separators.

> ⓘ The *system code* typically also simulates the *control logic* employed in the plant and is able to predict the evolution of *accidents*.

system code validation

See *validation* (1).

system code verification

See *verification* (1).

system for nuclear material accounting and control

An integrated set of measures designed to provide information on, control of and assurance of the presence of *nuclear material*, including those systems necessary to establish and track nuclear material inventories, control access to and detect loss or diversion of *nuclear material*, and ensure the *integrity* of those systems and measures.

> ⓘ For safeguards purposes, see the definition of State (or regional) system of accounting for and control of *nuclear material* in the Safeguards Glossary [14].

See also *control (of nuclear material)*.

system validation

See *validation* (2).

T

tailings

The residues resulting from the processing of ore to extract *uranium series* or *thorium series* radionuclides, or similar residues from processing ores for other purposes.

tank

A portable *tank* (including a *tank* container), a road *tank vehicle*, a rail *tank* wagon or a receptacle that contains solids, liquids, or gases, having a capacity of not less than 450 L when used for the transport of gases. (See SSR-6 (Rev. 1) [2].)

> ! This usage is specific to the Transport Regulations [2], and should otherwise be avoided.

target

Nuclear material, other radioactive material, associated facilities, associated activities, or other locations or objects of potential exploitation by a *nuclear security threat*, including *major public events, strategic locations, sensitive information,* and *sensitive information assets.*

> ⓘ In most cases, this should not need definition, as the general meaning is consistent with the normal dictionary meaning and the context should make clear the specific meaning.

target tissue or organ

The tissue or organ to which *radiation* is directed or the radiosensitive tissue or organ for which *dose* is assessed.

> ⓘ Used in internal dosimetry, normally in relation to a *source region.*

task related monitoring

See *monitoring* (1).

technological obsolescence

See *ageing*: *non-physical ageing.*

temporary relocation

See *relocation.*

therapeutic exposure

See *exposure categories*: *medical exposure.*

thermodynamic diameter

See *activity median aerodynamic diameter (AMAD).*

thorium series

The decay chain of ^{232}Th.

> ⓘ Namely, ^{232}Th, ^{228}Ra, ^{228}Ac, ^{228}Th, ^{224}Ra, ^{220}Rn, ^{216}Po, ^{212}Pb, ^{212}Bi, ^{212}Po (64%), ^{208}Tl (36%) and (stable) ^{208}Pb.

[thoron]

Radon-220.

> ! This usage is discontinued in the IAEA *safety standards* and should be avoided.

[thoron progeny]

The (short lived) *radioactive* decay products of ^{220}Rn.

> ! This usage is discontinued in the IAEA *safety standards* and should be avoided.

> ⓘ Namely, ^{216}Po (sometimes called thorium A), ^{212}Pb (thorium B), ^{212}Bi (thorium C), ^{212}Po (thorium C′, 64%) and ^{208}Tl (thorium C″, 36%). The stable decay product ^{208}Pb is sometimes known as thorium D.

threat

A person or group of persons with motivation, intention and capability to commit a *malicious act*.

> ⓘ In this usage, a *threat* is generally understood to be a postulated person or group against whose capabilities and intentions nuclear security measures are designed, whereas a real person or group who actually takes action to attempt a *malicious act* becomes an *adversary*. However, this distinction is not always maintained consistently, and in some cases it may be difficult to decide which term is more appropriate.

> ***design basis threat (DBT).*** The attributes and characteristics of potential *insider* and/or *external adversaries*, who might attempt unauthorized removal or *sabotage*, against which a *physical protection system* is designed and evaluated.

> ***representative threat statement.*** The attributes and characteristics of potential *insider* and/or *external adversaries* who might attempt *unauthorized removal* or *sabotage*, intended to be used to develop prescriptive requirements for the protection of defined materials and/or facilities.

> ***threat assessment.*** See *assessment* (1).

> ***threat statement.*** A description of credible adversaries (including attributes and characteristics) in the form of *design basis threat* or *representative threat statement*, developed on the basis of the national nuclear security *threat assessment*.

See also *adversary*.

through or into

Through or into the countries in which a *consignment* is transported but specifically excluding countries over which a *consignment* is carried by air, provided that there are no scheduled stops in those countries (see SSR-6 (Rev. 1) [2]).

> ! This usage is specific to the Transport Regulations [2], and should otherwise be avoided.

time based maintenance

See *maintenance*: *periodic maintenance*.

tissue equivalent material

Material designed to have, when irradiated, interaction properties similar to those of soft tissue.

ⓘ Used to make phantoms, such as the *ICRU sphere*.

ⓘ The *tissue equivalent material* used in the *ICRU sphere* has a density of 1 g/cm^3 and an elemental composition, by mass, of 76.2% oxygen, 11.1% carbon, 10.1% hydrogen and 2.6% nitrogen, but materials of various other compositions (e.g. water) are considered suitable for particular applications [30].

The term ***tissue substitute*** is also used with the same meaning.

tissue substitute

See *tissue equivalent material*.

tissue weighting factor, w_T

Multiplier of the *equivalent dose* to a tissue or organ used for purposes of *radiation protection* to account for the different sensitivities of different tissues or organs to the induction of *stochastic effects* of *radiation* [33].

Recommended *tissue weighting factors* for calculating *effective dose* are given in Table 5.

TABLE 5. TISSUE WEIGHTING FACTORS RECOMMENDED IN REF. [33]

Tissue or organ	w_T	$\sum w_T$
Bone-marrow (red), colon, lung, stomach, breast, remainder tissues[a]	0.12	0.72
Gonads	0.08	0.08
Bladder, oesophagus, liver, thyroid	0.04	0.16
Bone surface, brain, salivary glands, skin	0.01	0.04
Total		1.00

[a] The w_T for remainder tissues (0.12) applies to the arithmetic mean dose to these 13 organs and tissues for each sex: adrenals, extrathoracic (ET) region, gall bladder, heart, kidneys, lymphatic nodes, muscle, oral mucosa, pancreas, prostate (male), small intestine, spleen, thymus, uterus/cervix (female).

trace element

An element in a sample that has an average concentration of less than 1000 µg/g or 0.1% of the matrix composition.

ⓘ This term is used in various areas of science with different numerical meanings. This definition is specific to *nuclear forensics*.

transboundary exposure

See *exposure* (1).

transboundary movement

1.   Any movement of *radioactive material* from one State *through or into* another.

2.   [Any shipment of *spent fuel* or of *radioactive waste* from a State of origin to a State of destination.] (See Ref. [11].)

transient population group

Those *members of the public* who are residing for a short period of time (days to weeks) in a location (such as a camping site) that can be identified in advance. This does not include *members of the public* who may be travelling through an area.

transnational emergency

See *emergency*.

transport

1.   The deliberate physical movement of *radioactive material* (other than that forming part of the means of propulsion) from one place to another.

> ⓘ The term ***transportation*** is also used, in particular in US English or where there is a need to distinguish this meaning of *transport* from meaning (2).

> ⓘ In the context of the Transport Regulations, *transport* comprises all operations and conditions associated with, and involved in, the movement of *radioactive material*; these include the *design*, manufacture, maintenance and repair of *packaging*, and the preparation, consigning, loading, carriage including in-transit storage, *shipment* after storage, unloading and receipt at the final destination of loads of *radioactive material* and *packages*.

> ⓘ In some Nuclear Security Series publications addressing transport of nuclear material, *transport* has also been defined as "International or domestic carriage of *nuclear material* by any means of transport, beginning with the departure from a *nuclear facility* of the *shipper* and ending with the arrival at a *nuclear facility* of the receiver" [6].

> *[international nuclear transport].* The carriage of a consignment of nuclear material by any means of transportation intended to go beyond the territory of the State where the shipment originates, beginning with the departure from a facility of the shipper in that State and ending with the arrival at a facility of the receiver within the State of ultimate destination. (See Refs [4–6].)

> > ⓘ The 2005 Amendment to the Convention on the Physical Protection of Nuclear Material and Nuclear Facilities was adopted on 8 July 2005.

> > ⓘ More recent texts use the term *transboundary movement* for a similar concept.

2.   The movement of something as a result of being carried by a medium.

> ⓘ A general term used when a number of different *processes* are involved. The most common examples are heat *transport* — a combination of *advection*, convection, etc., in a cooling medium — and radionuclide *transport* in the *environment* — which could include *processes* such as *advection*, *diffusion*, *sorption* and *uptake*.

transport control centre

A facility which provides for the continuous monitoring of a *transport conveyance* location and security status and for communication with the *transport conveyance*, *shipper*/receiver, *carrier* and, when appropriate, its *guards,* and the *response forces*.

ⓘ This definition is for use in the context of *nuclear security*.

transport index (TI)

A number assigned to a *package, overpack* or *freight container*, or to unpackaged LSA-I or SCO-I or SCO-III, that is used to provide *control* over *radiation exposure*. (See SSR-6 (Rev. 1) [2].)

ⓘ The value of the *transport index* for a *package* or *overpack* is used (together with the surface *dose rate*) in determining the category (I-WHITE, II-YELLOW or III-YELLOW) to which the *package* or *overpack* belongs.

ⓘ A *package* or *overpack* with a *transport index* higher than 10 can be transported only under *exclusive use*.

ⓘ The *procedure* for calculating a *transport index* is given in section V of the Transport Regulations [2].

ⓘ In essence, the *transport index* is the maximum *dose rate* at 1 m from the outer surface of the load, expressed in mrem/h (or the value in mSv/h multiplied by 100), and in specified cases multiplied by a factor between 1 (for small sized loads) and 10 (for large sized loads). (See SSR-6 (Rev. 1) [2].)

transportation

See *transport* (1).

treatment

See *radioactive waste management* (1).

two-person rule

A procedure that requires at least two authorized and knowledgeable persons to be present to verify that activities involving *nuclear material* and *nuclear facilities* are authorized in order to detect access or actions that are unauthorized.

Type A/B(U)/B(M)/C package

See *package*.

type test

Conformity test made on one or more items representative of the production.

U

ultimate heat sink

A medium into which the transferred *residual heat* can always be accepted, even if all other means of removing the heat have been lost or are insufficient.

ⓘ This medium is normally a body of water or the atmosphere.

ultimate heat transport system

The *systems* and *components* needed to transfer *residual heat* to the *ultimate heat sink* after *shutdown*.

unacceptable radiological consequences

A level of radiological consequences, established by the State, above which the implementation of *nuclear security measures* is warranted.

! This definition is for use in the context of planning *nuclear security measures*.

unattached fraction

The fraction of *potential alpha energy* of *radon* decay products that arises from atoms that are not attached to ambient aerosol particles.

unauthorized act

See *malicious act*.

unauthorized removal

The theft or other unlawful taking of *radioactive material*.

uncertainty

aleatory uncertainty. *Uncertainty* inherent in a phenomenon.

ⓘ *Aleatory uncertainty* (or stochastic *uncertainty*) is taken into account by representing a phenomenon in terms of a probability distribution *model*.

ⓘ *Aleatory uncertainty* is of relevance for *events* or phenomena that occur in a random manner, such as random *failures* of items of equipment [19].

epistemic uncertainty. *Uncertainty* attributable to incomplete knowledge about a phenomenon, which affects the ability to model it.

ⓘ *Epistemic uncertainty* is reflected in a range of viable *models*, multiple expert interpretations and statistical confidence.

ⓘ *Epistemic uncertainty* is associated with the state of knowledge relating to a given problem under consideration. In any analysis or analytical *model* of a physical phenomenon, simplifications and assumptions are made. Even for relatively simple problems, a *model* may omit some aspects that are deemed unimportant to the solution.

ⓘ Additionally, the state of knowledge within the relevant scientific and engineering disciplines may be incomplete. Simplifications and incompleteness of knowledge give rise to *uncertainties* in the prediction of outcomes for a specified problem.

uncertainty analysis

See *analysis*.

uniform hazard response spectrum

See *response spectrum*.

unilateral approval

See *approval*.

unirradiated

> **unirradiated material.** Material not irradiated in a reactor or material irradiated in a reactor but with a radiation level equal to or less than 1 Gy/h (100 rad/h) at 1 m unshielded.
>
> ⓘ This definition appears as a footnote to Table 2. See *nuclear material*.
>
> **unirradiated thorium.** Thorium containing not more than 10^{-7} g of ^{233}U per gram of ^{232}Th. (See SSR-6 (Rev. 1) [2].)
>
> ⓘ Although the term *unirradiated thorium* is used, the issue is not really whether the thorium has been irradiated, but rather whether the content of ^{233}U (a *fissile material*) is significantly higher than the trace levels found in naturally occurring thorium.
>
> ! This usage is specific to the Transport Regulations [2].
>
> **unirradiated uranium.** Uranium containing not more than 2×10^3 Bq of plutonium per gram of ^{235}U, not more than 9×10^6 Bq of fission products per gram of ^{235}U and not more than 5×10^{-3} g of ^{236}U per gram of ^{235}U. (See SSR-6 (Rev. 1) [2].)
>
> ⓘ Although the term *unirradiated uranium* is used, the issue is not really whether the *uranium* has been irradiated, but rather whether the content of plutonium (a *fissile material*) is significantly higher than the trace levels found in naturally occurring *uranium*.
>
> ! This usage is specific to the Transport Regulations [2].

unrestricted linear energy transfer, L_∞

See *linear energy transfer (LET)*.

unrestricted use

See *use*.

unsealed source

See *source* (2).

unwitting insider

See *insider*.

uptake

1.   A general term for the *processes* by which radionuclides enter one part of a biological system from another.

> ⓘ Used for a range of situations, in particular for describing the overall effect when there are a number of contributing *processes*; for example, **root uptake**, the transfer of radionuclides from soil to plants through the plant roots.

2.   The *processes* by which radionuclides enter the body fluids from the respiratory tract, gastrointestinal tract or through the skin, or the fraction of an *intake* that enters the body fluids by these *processes*.

> ⓘ Also, the amount of material transferred from the site of *intake* to body organs or tissues.

uranium

> ***depleted uranium.*** *Uranium* containing a lesser mass percentage of ^{235}U than *natural uranium*. (See SSR-6 (Rev. 1) [2].)
>
> > ! This usage is specific to the Transport Regulations [2].
>
> ***enriched uranium.*** *Uranium* containing a higher mass percentage of ^{235}U than 0.72%. In all cases, a very small mass percentage of ^{234}U is present. (See SSR-6 (Rev. 1) [2].)
>
> > ! This usage is specific to the Transport Regulations [2].
>
> ***high enriched uranium (HEU).*** *Enriched uranium* containing 20% or more in weight percent of the isotope ^{235}U. HEU is considered a *special fissionable material* and a direct use material. (See Ref. [14].)
>
> ***low enriched uranium (LEU).*** *Enriched uranium* containing less than 20% in weight percent of the isotope ^{235}U. LEU is considered a *special fissionable material* and an indirect use material. (See Ref. [14].)
>
> ***natural uranium.*** *Uranium* (which may be chemically separated) containing the naturally occurring distribution of *uranium* isotopes (approximately 99.28% ^{238}U and 0.72% ^{235}U, by mass). (See SSR-6 (Rev. 1) [2].)
>
> > ! This usage is specific to the Transport Regulations [2].
>
> > ⓘ In all cases, a very small mass percentage of ^{234}U is present.
>
> > ⓘ The naturally occurring distribution of *uranium* isotopes including ^{234}U (approximately 99.285% ^{238}U, 0.710% ^{235}U, and 0.005% ^{234}U by mass) corresponds to approximately 48.9% ^{234}U, 2.2% ^{235}U and 48.9% ^{238}U by *activity*.

uranium enriched in the isotope uranium-235 or uranium-233

Uranium containing the isotope ^{235}U or ^{233}U or both in an amount such that the abundance ratio of the sum of these isotopes to the isotope ^{238}U is higher than the ratio of the isotope ^{235}U to the isotope ^{238}U occurring in nature (See Refs [4–6, 47]).

ⓘ The 2005 Amendment to the Convention on the Physical Protection of Nuclear Material and Nuclear Facilities was adopted on 8 July 2005.

uranium series

The decay chain of ^{238}U.

ⓘ Namely, ^{238}U, ^{234}Th, ^{234}Pa, ^{234}U, ^{230}Th, ^{226}Ra, ^{222}Rn, ^{218}Po, ^{214}Pb, ^{214}Bi, and ^{214}Po, ^{210}Pb, ^{210}Bi, ^{210}Po and (stable) ^{206}Pb, plus traces of ^{218}At, ^{210}Tl, ^{209}Pb, ^{206}Hg and ^{206}Tl.

urgent protective action

See *protective action* (1).

urgent protective action planning zone (UPZ)

See *emergency planning zone*.

urgent response phase

See *emergency response phase*.

use

authorized use. Use of *radioactive material* or *radioactive* objects from an *authorized facility or activity* in accordance with an *authorization*.

ⓘ Intended primarily for contrast with *clearance*, in that *clearance* implies no further *regulatory control* over the use, whereas the *authorization* for *authorized use* may prescribe or prohibit specific uses.

ⓘ A form of *restricted use*.

restricted use. The use of an *area* or of materials subject to restrictions imposed for reasons of *radiation protection and safety*.

ⓘ Restrictions would typically be expressed in the form of prohibition of particular *activities* (e.g. house building, growing or harvesting particular *foods*) or prescription of particular *procedures* (e.g. materials may only be recycled or reused within a *facility*).

unrestricted use. The use of an *area* or of material without any radiologically based restrictions.

! There may be other restrictions on the use of the *area* or material, such as planning restrictions on the use of an *area* of land or restrictions related to the chemical properties of a material.

! In some situations, these restrictions could, in addition to their primary intended effect, have an incidental effect on *radiation exposure*, but the use is classified as *unrestricted use* unless the primary reason for the restrictions is radiological.

ⓘ *Unrestricted use* is contrasted with *restricted use*.

validation

1.   The *process* of determining whether a product or service is adequate to perform its intended function satisfactorily.

⓵ *Validation* (typically of a *system*) concerns checking against the specification of *requirements*, whereas *verification* (typically of a design specification, a test specification or a test report) relates to the outcome of a *process*.

⓵ *Validation* may involve a greater element of judgement than *verification*.

model validation. The *process* of determining whether a *model* is an adequate representation of the real *system* being modelled, by comparing the predictions of the *model* with observations of the real *system*.

⓵ Usually contrasted with *model verification*, although *verification* will often be a part of the broader *process* of *validation*.

⓵ Modelling the behaviour of an engineered *system* in a geological *disposal facility* involves temporal scales and spatial scales for which no comparisons with system level tests are possible: *models* cannot be 'validated' for that which cannot be observed.

⓵ '*Model validation*' in these circumstances implies showing that there is a basis for confidence in the *model(s)* by means of detailed external reviews and comparisons with appropriate field and laboratory tests, and comparisons with observations of tests and of analogous materials, conditions and geologies at the *process* level.

⓵ What is typically required by *regulatory bodies* is that such *models* of the behaviour of engineered systems in a geological *disposal facility* be shown to be 'fit for purpose'; this is typically called '*validation*' in national regulations.

system code validation. *Assessment* of the *accuracy* of values predicted by the *system code* against relevant experimental data for the important phenomena expected to occur.

accuracy. In this context, the known bias between the prediction of a *system code* and the actual performance in transients of a *facility*.

2.   Confirmation by *examination* and by means of objective evidence that specified objectives have been met and specified *requirements* for a specific intended purpose and use or application have been fulfilled.

See also *verification*.

⓵ The corresponding status is termed 'validated'.

⓵ *Validation* typically entails the *assessment* of a final product against its specified objectives and specified *requirements*.

⓵ The conditions of use for *validation* purposes may be real or simulated.

system validation. Confirmation by *examination* and provision of evidence that a *system* fulfils in its entirety the specification of *requirements* as intended (e.g. *validation* of an instrumentation and *control system* in terms of functionality, response time, fault tolerance and robustness).

3.   A means of *multilateral approval* of a *transport package design* or *shipment*, whereby an endorsement on the original *certificate* or the issuance of a separate endorsement, annex, supplement, etc., is produced by the *competent authority* of the country *through or into* which the *shipment* is made. (See SSR-6 (Rev. 1) [2].)

vehicle

A road *vehicle* (including an articulated *vehicle*, i.e. a tractor and semi-trailer combination), railroad car or railway wagon. Each trailer shall be considered a separate *vehicle*. (See SSR-6 (Rev. 1) [2].)

! This usage is specific to the Transport Regulations [2], and should otherwise be avoided.

vendor

A *design*, contracting or manufacturing organization supplying a service, *component* or *facility*.

vent

An opening in the *Earth's crust* where volcanic products (e.g. *lava*, solid rock, gas, liquid water) are erupted.

ⓘ *Vents* may be either circular structures (i.e. craters) or elongate fissures or fractures, or small cracks in the ground.

verification

1.   The *process* of determining whether the quality or performance of a product or service is as stated, as intended or as required.

ⓘ *Verification* is closely related to quality management and *quality control*.

model verification. The *process* of determining whether a *computational model* correctly implements the intended *conceptual model* or *mathematical model*.

system code verification. Review of source coding in relation to its description in the *system code* documentation.

See also *site evaluation*: *site verification*.

2.   Confirmation by *examination* and by means of objective evidence that specified objectives have been met and specified *requirements* for specific results have been fulfilled.

ⓘ The corresponding status is termed 'verified'.

ⓘ *Verification* typically entails the *assessment* of the results of an individual activity against its inputs.

ⓘ *Verification* may comprise *activities* such as: performing alternative calculations; comparing a new *design* specification with a similar proven *design* specification; undertaking tests and demonstrations; and reviewing documents prior to issue.

See also *validation*.

very low level waste (VLLW)

See *waste classes*.

very short lived waste

See *waste classes*.

vessel (for carrying cargo)

Any sea-going *vessel* or inland waterway craft used for carrying cargo. (See SSR-6 (Rev. 1) [2].)

> ! This restrictive use of the term *vessel* in relation to the *transport* of *radioactive material* does not apply in other areas of *safety*; for example, a reactor pressure vessel is a vessel as usually understood.

vital area

See *area*.

volcanic activity

A feature or process on a *volcano* or within a *volcanic field* that is linked to the presence of *magma* and heat gases emanating from the Earth and their interaction with nearby crustal rocks or groundwater.

> ⓘ *Volcanic activity* includes seismicity, fumarolic activity, high rates of heat flow, emission of ground gases, thermal springs, deformation, ground cracks, pressurization of aquifers and ash venting. The term includes *volcanic unrest* and *volcanic eruption*.

volcanic earthquake

A seismic event caused by, and directly associated with, processes in a *volcano*.

> ⓘ *Volcanic earthquakes* and seismic activity come in many forms and types (e.g. *volcano*–tectonic earthquakes, long period events, hybrid events, tremors, swarms) before, during and after *volcanic eruption*s, and their characteristics and patterns are used to infer what is happening within the *volcano* at different times.

> ⓘ Seismic monitoring is the most fundamental method used for forecasting the onset of a *volcanic eruption* and for assessing the potential for *volcanic eruption*.

> ⓘ Increasing seismicity, continuous tremor, shift in *hypocentres* towards the surface with time and the occurrence of shallow long period (or low frequency) events imply a high possibility that the onset of *volcanic eruption* is very close. Tremors can also continue through *volcanic eruptions*.

volcanic eruption

Any process on a *volcano* or at a *volcanic vent* that involves the explosive ejection of fragmental material, the effusion of molten *lava*, the sudden release of large quantities of volcanic gases (e.g. CO_2) or a process by which buried regions of the volcanic systems from various depths, such as the hydrothermal system, are brought to the surface during the collapse of edifices.

> ⓘ *Volcanic eruptions* are magmatic if newly solidified *magma* is present in the eruptive products and non-magmatic (phreatic) if they involve only recycled rock fragments. *Volcanic eruptions* can occur over widely varying timescales (seconds to years).

> *effusive eruption.* A *volcanic eruption* in which coherent *magma* is extruded from the *volcanic vent* to form *lava* flows.

> *explosive eruption.* A *volcanic eruption* in which gas bubble expansion or explosive interaction between *magma* and water is rapid enough to break the *magma* apart (i.e. to fragment the *magma*).

> ⓘ *Explosive eruptions* also occur when pressurized hydrothermal gases and superheated fluids suddenly break the host rock in a volcanic edifice.

ⓘ Pyroclastic flows, falls and *volcano* generated missiles are characteristic of *explosive eruptions*.

phreatic eruption. A type of *eruption* caused by rapid volume expansion of water, or water vaporization, in the subsurface, without *magma* being erupted at the surface.

ⓘ *Phreatic eruptions* are usually steam explosions that occur when hot water is suddenly depressurized, but may occasionally be non-explosive expulsions of pressurized or heated aquifer waters and/or hydrothermal fluids at a *volcano*.

ⓘ *Phreatic eruptions* are common where rising *magma* interacts with groundwater, commonly in the interior of a *volcano* edifice.

ⓘ Although commonly small in scale, *phreatic eruptions* may be followed by larger scale *phreatomagmatic eruptions* or magmatic *eruptions*.

ⓘ *Phreatic eruptions* may generate debris flows and hot lahars.

phreatomagmatic eruption. A type of *explosive eruption* that involves subsurface interaction of *magma* and water and which produces explosive mixtures of rock, steam and *magma* that often form pyroclastic flows and surges.

ⓘ Surtseyan and phreato-plinian eruptions are *phreatomagmatic eruptions* involving the interaction of hot pyroclasts and water, as the *magma* is erupted from the *volcanic vent* into bodies of water.

plinian eruption. An explosive pyroclastic eruption characterized by a sustained eruption column that generally rises to altitudes of 10–50 km.

ⓘ *Plinian eruptions* may produce thick tephra fallout over areas of 500–5000 km^2 and/or pyroclastic flows and surges that travel tens of kilometres from the *volcano*.

ⓘ The 1991 eruption of Mount Pinatubo, Philippines, is a recent *plinian eruption*.

strombolian eruption. A type of *volcanic eruption* that is intermediate in explosivity between fire fountain and *plinian eruptions*.

ⓘ *Magma* is less fragmented in a *strombolian eruption* than in a *plinian eruption* and gas is often released in coalesced slugs rather than in a continuous jet.

ⓘ *Strombolian eruptions* are commonly discrete events, punctuated by intervals of relative quiescence lasting from a few seconds to several hours.

ⓘ *Strombolian eruptions*, usually basaltic to andesitic in composition, form weak eruption columns that rarely exceed 5 km in height, and the volume of *lava* flows is generally equal to, or greater than, the volume of pyroclastic rocks.

ⓘ Such eruptions are characteristic of Stromboli *volcano*, Italy, and Izalco *volcano*, El Salvador.

vulcanian eruption. A type of *volcanic eruption* characterized by discrete explosions, which produces shock waves and pyroclastic eruptions.

ⓘ *Vulcanian eruptions* typically occur when volcanic gas accumulates in a solidifying shallow conduit or dome, which pressurizes the *magma* to the point of brittle *failure*.

ⓘ Andesitic and dacitic *magmas* are most often associated with *vulcanian eruptions*.

ⓘ Examples of recent *vulcanian eruptions* include Sakurajima *volcano*, Japan, Soufrière Hills *volcano*, Montserrat, and Colima *volcano*, Mexico.

volcanic event

Any occurrence, or sequence of phenomena, associated with *volcanoes* that may give rise to *volcanic hazards*.

ⓘ *Volcanic events* may be formally defined as part of a *hazard assessment* in order to provide meaningful definition of repose intervals and *hazards*.

ⓘ *Volcanic events* may include *volcanic eruptions* and will typically include the occurrence of non-eruptive *hazards*, such as landslides.

volcanic field

Any spatial cluster of *volcanoes*.

Also termed **volcano group**.

ⓘ *Volcanic fields* range in size from a few volcanoes to over 1000 volcanoes.

ⓘ *Volcanic fields* may consist of monogenetic *volcanoes* (e.g. the Cima *volcanic field*, United States of America), or both polygenetic and monogenetic *volcanoes* (e.g. the Kluchevskoy *volcano group*, Russian Federation).

volcanic hazard

A volcanic process or phenomenon that can have an adverse effect on people or infrastructure.

ⓘ In the more restricted context of *risk assessment*, it is the probability of occurrence, within a specific period of time in a given *area*, of a potentially damaging *volcanic event* of a given intensity value (e.g. thickness of tephra fallout).

volcanic unrest

Variation in the nature, intensity, spatio-temporal distribution and chronology of geophysical, geochemical and geological activity and phenomena as observed and recorded on a *volcano*, from a baseline level of activity known for this *volcano* or for other similar *volcanoes* outside periods of eruptive activity.

ⓘ *Volcanic unrest* can be precursory and can culminate in a *volcanic eruption*, although in most cases, rising *magma* or pressurized fluids that cause unrest do not breach the surface and erupt.

volcanic vent

See *vent*.

volcano

A naturally occurring *vent* at the Earth's surface through which *lava*, solid rock and associated gases and liquid water can erupt.

ⓘ The edifice that is built by the explosive or effusive accumulation of these products over time is also a *volcano*.

capable volcano. A *volcano* that has a credible likelihood of undergoing future activity and producing hazardous phenomena, including non-eruptive phenomena, during the *lifetime* of a *nuclear installation* concerned, and which may potentially affect the site.

ⓘ Hierarchical criteria for determining whether a *volcano* or *volcanic field* is a *capable volcano* or a **capable volcanic field** are: (i) evidence of contemporary volcanic activity or active near surface processes associated with magnetism for any *volcano* in the geographical region; (ii) Holocene volcanic activity for any *volcano* within the geographical region; and (iii) some evidence of potential for activity, such as recurrence rates of volcanism greater than 10^{-7} per year, and the potential to produce hazardous phenomena that may affect the site vicinity [58].

Holocene volcano. A *volcano* or *volcanic field* that has erupted within the past 10 000 years (the *Holocene*).

ⓘ Reported historical activity and radiometric dating of volcanic products provide the most direct evidence of *volcanic eruptions* within the *Holocene*.

ⓘ In some circumstances, especially in the early stages of site investigation, the exact age of the most recent volcanic products may be difficult to determine.

ⓘ In such circumstances, additional evidence may be used to judge a *volcano* as *Holocene* (e.g. by following the methods used by the Smithsonian Institution, United States of America).

ⓘ Such evidence includes: (i) volcanic products overlying latest Pleistocene glacial debris; (ii) youthful volcanic landforms in areas where erosion would be expected to be pronounced after many thousands of years; (iii) vegetation patterns that would have been far richer if the volcanic substrates were more than a few thousand (or hundred) years old; and (iv) ongoing fumarolic degassing, or the presence of a hydrothermal system at the *volcano*.

ⓘ In addition, some *volcanoes* may be denoted as Holocene(?) volcanoes if authorities disagree over the existence of *Holocene* volcanism, or when the original investigator expresses uncertainty about the most reliable age estimate of the most recent *volcanic eruption*.

ⓘ Under these circumstances, it is reasonable to consider such *volcanoes* to be *Holocene* and to proceed with the *hazard assessment*.

volcano explosivity index (VEI)

A classification scheme for the explosive magnitude of a *volcanic eruption*, primarily defined in terms of the total volume of erupted tephra, but in some cases the height of the eruption column and the duration of continuous *explosive eruption* are used to determine the VEI value.

ⓘ The *VEI* varies from VEI 0 (non-*explosive eruption*, less than 10^4 m^3 tephra ejected) to VEI 8 (largest *explosive eruption* identified in the geological record, more than 10^{12} m^3 tephra ejected).

ⓘ A unit of increasing explosivity on the *VEI* scale generally corresponds to an increase in volume of erupted tephra by a factor of ten.

ⓘ The only exception is the transition from VEI 0 to VEI 1, which represents an increase in the volume of tephra erupted by a factor of one hundred.

volcano generated missile

A pyroclastic particle, often of large size, that is forcefully ejected, follows a high angle trajectory from the *volcanic vent* to the surface as a result of explosive activity at the *vent* and falls under gravity.

ⓘ *Volcano generated missiles* can be of any material, such as rock fragments, trees and structural debris, that is rapidly transported by flow phenomena with significant momentum and that may impact *structures*, causing considerable damage, even beyond the extent of the main flow itself.

volcano group

See volcanic field.

volcano monitoring

Geophysical, geochemical and geological monitoring to evaluate the potential for a forthcoming *volcanic eruption*, forecast the onset of *eruption*, understand an ongoing *eruption* and evaluate the potential *volcanic hazards* arising from an *eruption*.

ⓘ Instruments such as seismometers, global positioning system receivers, tiltmeters, magnetometers, gas sensors, cameras and/or related instruments are installed on and around the *volcano* to evaluate *volcanic activity*, identify *volcanic unrest* and evaluate the potential for *volcanic eruption*.

ⓘ Remote sensing by satellite is sometimes very effective in monitoring temporal thermal, topographical and geological changes in *volcanoes*.

volume reduction

See *radioactive waste management* (1).

vulcanian eruption

See volcanic *eruption*.

vulnerable source

See *source* (2).

vulnerability

A physical feature or operational attribute that renders an entity, asset, system, network, facility, activity or geographic area open to exploitation or susceptible to a given *threat*.

> ***vulnerability assessment.*** See *assessment* (1).

W

warning point

A designated organization to act as a point of contact that is staffed or able to be alerted at all times for promptly responding to, or initiating a response to, an incoming *notification* (definition (2)), warning message, request for assistance or request for *verification* of a message, as appropriate, from the IAEA.

waste

Material for which no further use is foreseen.

> ***exempt waste.*** *Waste* from which *regulatory control* is removed in accordance with *exemption* principles.
>
> ⓘ This is *waste* that meets the criteria for *clearance*, *exemption* or *exclusion* from *regulatory control* for *radiation protection* purposes as described in Refs [20, 59].
>
> ! This is therefore not *radioactive waste*.
>
> **[*mining and milling waste (MMW)*].** *Waste* from *mining and milling.*
>
> ⓘ This includes *tailings* from *processing*, residues from heap leaching, waste rock, sludges, filter cakes, scales and various effluents.

See also [*mining and milling*].

> ***mixed waste.*** *Radioactive waste* that also contains non-*radioactive* toxic or hazardous substances.
>
> ***NORM waste.*** *Naturally occurring radioactive material (NORM)* for which no further use is foreseen.
>
> ***secondary waste.*** *Radioactive waste* resulting as a byproduct from the processing of primary *radioactive waste*.

See also *radioactive waste*.

waste, radioactive

See *radioactive waste*.

waste acceptance criteria

Quantitative or qualitative criteria specified by the *regulatory body*, or specified by an *operator* and approved by the *regulatory body*, for the *waste form* and *waste package* to be accepted by the *operator* of a *waste management facility*.

> ⓘ *Waste acceptance criteria* specify the radiological, mechanical, physical, chemical and biological characteristics of *waste packages* and unpackaged *waste*.
>
> ⓘ *Waste acceptance criteria* might include, for example, restrictions on the *activity concentration* or total *activity* of particular radionuclides (or types of radionuclide) in the *waste*, on their heat output or on the properties of the *waste form* or of the *waste package*.
>
> ⓘ *Waste acceptance criteria* are based on the *safety case* for the *facility* or are included in the *safety case* as part of the *operational limits and conditions* and controls.
>
> ⓘ *Waste acceptance criteria* are sometimes referred to as 'waste acceptance *requirements*'.

waste canister

See *waste container*.

waste characterization

See *characterization* (2).

waste classes

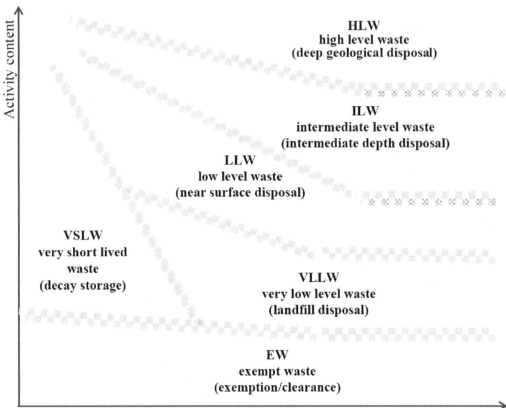

FIG. 6. Conceptual illustration of the waste classification scheme.

ⓘ *Waste classes* are those recommended in GSG-1 [59].

ⓘ This classification system is organized to take into account matters considered of prime importance for the *safety* of *disposal* of *radioactive waste*.

ⓘ The term 'activity content' is used because of the generally heterogeneous nature of *radioactive waste*; it is a generic term that covers *activity concentration*, *specific activity* and total *activity*.

ⓘ The other classes listed below (in square brackets) are sometimes used, for example in national classification systems, and are mentioned here to indicate how they typically relate to the classes in GSG-1 [59].

ⓘ Other systems classify *waste* on other bases, such as according to its origin (e.g. reactor operations *waste*, *reprocessing waste*, *decommissioning waste* and defence *waste*).

exempt waste. See *waste.*

[*heat generating waste (HGW)*]. *Radioactive waste* that is sufficiently *radioactive* that the decay heat significantly increases its temperature and the temperature of its surroundings.

ⓘ In practice, *heat generating waste* is usually *high level waste*, although some types of *intermediate level waste* may qualify as *heat generating waste*.

high level waste (HLW). The *radioactive* liquid containing most of the *fission products* and actinides present in *spent fuel* — which forms the residue from the first solvent extraction cycle in *reprocessing* — and some of the associated *waste* streams; this material following solidification; *spent fuel* (if it is declared as *waste*); or any other *waste* with similar radiological characteristics.

ⓘ Typical characteristics of *high level waste* are concentrations of long lived radionuclides exceeding the limitations for *short lived waste* [59].

ⓘ This is *waste* with levels of *activity concentration* high enough to generate significant quantities of heat by the *radioactive* decay process or *waste* with large amounts of long lived radionuclides that need to be considered in the design of a *disposal facility* for such *high level waste*.

ⓘ *Disposal* in deep, stable geological formations usually several hundred metres or more below the surface is the generally recognized option for the *disposal* of *high level waste*.

intermediate level waste (ILW). *Radioactive waste* that, because of its content, in particular its content of long lived radionuclides, requires a greater degree of *containment* and *isolation* than that provided by *near surface disposal*.

ⓘ Typical characteristics of *intermediate level waste* are levels of *activity concentration* above clearance levels.

ⓘ However, *intermediate level waste* needs no provision, or only limited provision, for heat dissipation during its *storage* and *disposal* [59].

ⓘ *Intermediate level waste* may contain long lived radionuclides, in particular, alpha emitting radionuclides that will not decay to a level of *activity concentration* acceptable for *near surface disposal* during the time for which *institutional controls* can be relied upon.

ⓘ *Waste* in this class may therefore require *disposal* at greater (intermediate) depths, of the order of tens of metres to a few hundred metres or more.

ⓘ *Intermediate level waste* may be so classified on the basis of *waste acceptance criteria* for *near surface disposal facilities.*

long lived waste. *Radioactive waste* that contains significant levels of radionuclides with a *half-life* greater than 30 years.

ⓘ Typical characteristics are long lived radionuclide concentrations exceeding the limitations for *short lived waste* [59].

low level waste (LLW). *Radioactive waste* that is above *clearance levels*, but with limited amounts of long lived radionuclides.

ⓘ *Low level waste* covers a very broad range of waste. Typical characteristics of *low level waste* are levels of *activity concentration* above clearance levels.

ⓘ *Low level waste* may include short lived radionuclides at higher levels of *activity concentration*, and also long lived radionuclides, but only at relatively low levels of *activity concentration* that require only the levels of *containment* and *isolation* provided by a *near surface disposal facility* [59].

227

ⓘ *Low level waste* requires robust *containment* and *isolation* for periods typically of up to a few hundred years and is suitable for *disposal* in engineered *near surface disposal facilities.*

ⓘ *Low level waste* may be so classified on the basis of *waste acceptance criteria* for *near surface disposal facilities.*

short lived waste. *Radioactive waste* that does not contain significant levels of radionuclides with a *half-life* greater than 30 years.

ⓘ Typical characteristics are restricted long lived radionuclide concentrations (limitation of long lived radionuclides to 4000 Bq/g in individual *waste packages* and to an overall average of 400 Bq/g per *waste package*); see para. 2.27 of GSG-1 [59].

very low level waste (VLLW). *Radioactive waste* that does not necessarily meet the criteria of *exempt waste*, but that does not need a high level of *containment* and *isolation* and, therefore, is suitable for *disposal* in landfill type near surface repositories with limited *regulatory control.*

ⓘ Such landfill type near surface repositories may also contain other hazardous waste; typical *waste* in this class includes soil and rubble with low levels of *activity concentration.*

ⓘ Concentrations of longer lived radionuclides in *very low level waste* are generally very limited [20, 59].

ⓘ This is a category used in some Member States; in others there is no such category, as no *radioactive waste* at all may be disposed of in this way, however low level it is.

very short lived waste. *Radioactive waste* that can be stored for decay over a limited period of up to a few years and subsequently cleared from *regulatory control* according to arrangements approved by the *regulatory body*, for uncontrolled *disposal*, use or *discharge* [20, 59].

ⓘ This class includes *radioactive waste* containing primarily radionuclides with very short half-lives often used for research and medical purposes.

waste conditioning

See *radioactive waste management* (1): *conditioning.*

waste container

The vessel into which the *waste form* is placed for handling, *transport*, *storage* and/or eventual *disposal*; also the outer *barrier* protecting the *waste* from external intrusions. The *waste container* is a *component* of the *waste package*. For example, molten *high level waste* glass would be poured into a specially designed *container* (*canister*), where it would cool and solidify.

! Note that the term **waste canister** is considered to be a specific term for a *container* for *spent fuel* or vitrified *high level waste.*

waste disposal

See *disposal.*

waste form

Waste in its physical and chemical form after *treatment* and/or *conditioning* (resulting in a solid product) prior to *packaging.*

ⓘ The *waste form* is a *component* of the *waste package.*

waste generator

The *operating organization* of a *facility or activity* that generates *waste*.

> ! For convenience, the scope of the term *waste generator* is sometimes extended to include whoever currently has the responsibilities of the *waste generator* (e.g. if the actual *waste generator* is unknown or no longer exists and a successor organization has assumed responsibility for the *waste*).

waste management, radioactive

See *radioactive waste management*.

waste management facility, radioactive

See *radioactive waste management facility*.

waste minimization

See *minimization of waste*

waste package

The product of *conditioning* that includes the *waste form* and any *container(s)* and internal *barriers* (e.g. absorbing materials and liner), as prepared in accordance with *requirements* for handling, *transport*, *storage* and/or *disposal*.

weakly penetrating radiation

See *radiation*.

wet storage

See *storage*.

worker

Any person who works, whether full time, part time or temporarily, for an *employer* and who has recognized rights and duties in relation to occupational *radiation protection*.

> ⓘ A self-employed person is regarded as having the duties of both an *employer* and a *worker*.

workers' health surveillance

Medical supervision intended to ensure the initial and continuing fitness of *workers* for their intended tasks.

[working level (WL)]

A unit of *potential alpha energy* concentration (i.e. the *potential alpha energy* per unit volume of air) resulting from the presence of decay products of ^{222}Rn or ^{220}Rn, equal to 1.3×10^8 MeV/m^3 (exactly).

> ! The term *working level* is now obsolete and its use is discouraged.

> ⓘ In SI units, a *working level* is 2.1×10^{-5} J/m^3 (approximately).

[working level month (WLM)]

The *exposure* due to decay products of ^{222}Rn or ^{220}Rn that would be incurred during a working month (170 hours) in a constant *potential alpha energy* concentration of one *working level*.

! The term *working level month* is now obsolete and its use is discouraged.

ⓘ In SI units, a *working level month* is 3.54×10^{-3} J·h/m^3 (approximately).

workplace monitoring

See *monitoring* (1).

REFERENCES

[1]   EUROPEAN COMMISSION, FOOD AND AGRICULTURE ORGANIZATION OF THE UNITED NATIONS, INTERNATIONAL ATOMIC ENERGY AGENCY, INTERNATIONAL LABOUR ORGANIZATION, OECD NUCLEAR ENERGY AGENCY, PAN AMERICAN HEALTH ORGANIZATION, UNITED NATIONS ENVIRONMENT PROGRAMME, WORLD HEALTH ORGANIZATION, Radiation Protection and Safety of Radiation Sources: International Basic Safety Standards, IAEA Safety Standards Series No. GSR Part 3, IAEA, Vienna (2014).

[2]   INTERNATIONAL ATOMIC ENERGY AGENCY, Regulations for the Safe Transport of Radioactive Material, 2018 Edition, IAEA Safety Standards Series No. SSR-6 (Rev. 1), IAEA, Vienna (2018).

[3]   INTERNATIONAL ATOMIC ENERGY AGENCY, Radioactive Waste Management Glossary, IAEA, Vienna (2003).

[4]   Convention on the Physical Protection of Nuclear Material, INFCIRC/274/Rev.1, IAEA, Vienna (1980).

[5]   Amendment to the Convention on the Physical Protection of Nuclear Material, INFCIRC/274/Rev. 1/Mod. 1 (Corrected), IAEA, Vienna (2021).

[6]   INTERNATIONAL ATOMIC ENERGY AGENCY, Nuclear Security Recommendations on Physical Protection of Nuclear Material and Nuclear Facilities, INFCIRC/225/Revision 5, IAEA Nuclear Security Series No. 13, IAEA, Vienna (2011).

[7]   INTERNATIONAL ATOMIC ENERGY AGENCY, Nuclear Security Recommendations on Radioactive Material and Associated Facilities, IAEA Nuclear Security Series No. 14, IAEA, Vienna (2011).

[8]   INTERNATIONAL ATOMIC ENERGY AGENCY, Nuclear Security Recommendations on Nuclear and Other Radioactive Material out of Regulatory Control, IAEA Nuclear Security Series No. 15, IAEA, Vienna (2011).

[9]   INTERNATIONAL ATOMIC ENERGY AGENCY, Objective and Essential Elements of a State's Nuclear Security Regime, IAEA Nuclear Security Series No. 20, IAEA, Vienna (2013).

[10] Convention on Nuclear Safety, INFCIRC/449, IAEA, Vienna (1994).

[11] Joint Convention on the Safety of Spent Fuel Management and on the Safety of Radioactive Waste Management, INFCIRC/546, IAEA, Vienna (1997).

[12] International Convention for the Suppression of Acts of Nuclear Terrorism, United Nations, New York (2005).

[13] INTERNATIONAL ATOMIC ENERGY AGENCY, Governmental, Legal and Regulatory Framework for Safety, IAEA Safety Standards Series No. GSR Part 1 (Rev. 1), IAEA, Vienna (2016).

[14] INTERNATIONAL ATOMIC ENERGY AGENCY, IAEA Safeguards Glossary (2022 Edition), IAEA, Vienna (in preparation).

[15] Convention on Early Notification of a Nuclear Accident, INFCIRC/335, IAEA, Vienna (1986).

[16] INTERNATIONAL ATOMIC ENERGY AGENCY, INES: The International Nuclear and Radiological Event Scale User's Manual, 2008 Edition, IAEA, Vienna (2013).

[17] INTERNATIONAL ORGANIZATION FOR STANDARDIZATION, Nuclear Energy: Vocabulary (Second Edition), ISO 921:1997, ISO, Geneva (1997).

[18] INTERNATIONAL COMMISSION ON RADIOLOGICAL PROTECTION, Optimization and Decision-making in Radiological Protection, Publication 55, Pergamon Press, Oxford and New York (1987).

[19] INTERNATIONAL ATOMIC ENERGY AGENCY, Safety Assessment for Facilities and Activities, IAEA Safety Standards Series No. GSR Part 4 (Rev. 1), IAEA, Vienna (2016).

[20] INTERNATIONAL ATOMIC ENERGY AGENCY, Application of the Concepts of Exclusion, Exemption and Clearance, IAEA Safety Standards Series No. RS-G-1.7, IAEA, Vienna (2004).

[21] INTERNATIONAL ATOMIC ENERGY AGENCY, Code of Conduct on the Safety and Security of Radioactive Sources, IAEA/CODEOC/2004, IAEA, Vienna (2004).

[22] FOOD AND AGRICULTURE ORGANIZATION OF THE UNITED NATIONS, INTERNATIONAL ATOMIC ENERGY AGENCY, INTERNATIONAL CIVIL AVIATION ORGANIZATION, INTERNATIONAL LABOUR ORGANIZATION, INTERNATIONAL MARITIME ORGANIZATION, INTERPOL, OECD NUCLEAR ENERGY AGENCY, PAN AMERICAN HEALTH ORGANIZATION, PREPARATORY COMMISSION FOR THE COMPREHENSIVE NUCLEAR-TEST-BAN TREATY ORGANIZATION, UNITED NATIONS ENVIRONMENT PROGRAMME, UNITED NATIONS OFFICE FOR THE COORDINATION OF HUMANITARIAN AFFAIRS, WORLD HEALTH ORGANIZATION, WORLD METEOROLOGICAL ORGANIZATION, Preparedness and Response for a Nuclear or Radiological Emergency, IAEA Safety Standards Series No. GSR Part 7, IAEA, Vienna (2015).

[23] Convention on the Prevention of Marine Pollution by Dumping of Wastes and Other Matter, International Maritime Organization, London (1972).

[24] EUROPEAN ATOMIC ENERGY COMMUNITY, INTERNATIONAL ATOMIC ENERGY AGENCY, FOOD AND AGRICULTURE ORGANIZATION OF THE UNITED NATIONS, INTERNATIONAL LABOUR ORGANIZATION, INTERNATIONAL MARITIME ORGANIZATION, OECD NUCLEAR ENERGY AGENCY, PAN AMERICAN HEALTH ORGANIZATION, UNITED NATIONS ENVIRONMENT PROGRAMME, WORLD HEALTH ORGANIZATION, Fundamental Safety Principles, IAEA Safety Standards Series No. SF-1, IAEA, Vienna (2006).

[25] INTERNATIONAL ATOMIC ENERGY AGENCY, Safety of Nuclear Power Plants: Design, IAEA Safety Standards Series No. SSR-2/1 (Rev. 1), IAEA, Vienna (2016).

[26] INTERNATIONAL NUCLEAR SAFETY ADVISORY GROUP, Defence in Depth in Nuclear Safety, INSAG Series No. 10, IAEA, Vienna (1996).

[27] INTERNATIONAL COMMISSION ON RADIOLOGICAL PROTECTION, Limits for Intakes of Radionuclides by Workers, Publication 30, Pergamon Press, Oxford and New York (1979–1982). (Partly superseded and supplemented by Refs [28] and [29].)

[28] INTERNATIONAL COMMISSION ON RADIOLOGICAL PROTECTION, Dose Coefficients for Intakes of Radionuclides by Workers, ICRP Publication No. 68, Ann. ICRP **24** 4, Elsevier Science, Oxford (1994).

[29] INTERNATIONAL COMMISSION ON RADIOLOGICAL PROTECTION, Age-dependent Doses to Members of the Public from Intakes of Radionuclides: Part 5, Compilation of Ingestion and Inhalation Dose Coefficients, Publication 72, Pergamon Press, Oxford and New York (1996).

[30] INTERNATIONAL COMMISSION ON RADIATION UNITS AND MEASUREMENTS, Quantities and Units in Radiation Protection Dosimetry, Rep. 51, ICRU, Bethesda, MD (1993).

[31] INTERNATIONAL COMMISSION ON RADIATION UNITS AND MEASUREMENTS, Fundamental Quantities and Units for Ionizing Radiation, Rep. 60, ICRU, Bethesda, MD (1998).

[32] INTERNATIONAL COMMISSION ON RADIATION UNITS AND MEASUREMENTS, Determination of Dose Equivalents Resulting from External Radiation Sources, Rep. 39, ICRU, Bethesda, MD (1985).

[33] INTERNATIONAL COMMISSION ON RADIOLOGICAL PROTECTION, The 2007 Recommendations of the International Commission on Radiological Protection, Publication 103, Ann. ICRP **37** 2–4, Elsevier Science, Oxford (2007).

[34] STEVENSON, A., WAITE, M. (Eds), Concise Oxford English Dictionary, 12th Edition, Oxford University Press, Oxford (2011).

[35] INTERNATIONAL COMMISSION ON RADIOLOGICAL PROTECTION, Guide for the Practical Application of the ICRP Human Respiratory Tract Model, ICRP Supporting Guidance 3, Ann. ICRP **32** 1–2 (2003).

[36] INTERNATIONAL ATOMIC ENERGY AGENCY, Safety of Research Reactors, IAEA Safety Standards Series No. SSR-3, IAEA, Vienna (2016).

[37] INTERNATIONAL ATOMIC ENERGY AGENCY, Safety of Nuclear Fuel Cycle Facilities, IAEA Safety Standards Series No. SSR-4, IAEA, Vienna (2017).

[38] INTERNATIONAL COMMISSION ON RADIOLOGICAL PROTECTION, Conversion Coefficients for Use in Radiological Protection against External Radiation, ICRP Publication 74, Ann. ICRP **26** 3, Pergamon Press, Oxford and New York (1997).

[39] INTERNATIONAL COMMISSION ON RADIOLOGICAL PROTECTION, Human Alimentary Tract Model for Radiological Protection, ICRP Publication No. 100, Ann. ICRP **36** 1–2, Elsevier Science, Oxford (2006).

[40] INTERNATIONAL ATOMIC ENERGY AGENCY, Seismic Hazards in Site Evaluation for Nuclear Installations, IAEA Safety Standards Series No. SSG-9 (Rev. 1), IAEA, Vienna (2022).

[41] INTERNATIONAL COMMISSION ON RADIATION UNITS AND MEASUREMENTS, Radiation Quantities and Units, Rep. 33, ICRU, Bethesda, MD (1980).

[42] Convention on Supplementary Compensation for Nuclear Damage, INFCIRC/567, IAEA, Vienna (1998).

[43] STOIBER, C., BAER, A., PELZER, N., TONHAUSER, W., Handbook on Nuclear Law, IAEA, Vienna (2003).

[44] INTERNATIONAL COMMISSION ON RADIOLOGICAL PROTECTION, 1990 Recommendations of the ICRP, Publication 60, Pergamon Press, Oxford and New York (1991).

[45] INTERNATIONAL COMMISSION ON RADIOLOGICAL PROTECTION, Age-dependent Doses to Members of the Public from Intakes of Radionuclides: Part 4, Inhalation Dose Coefficients, Publication 71, Pergamon Press, Oxford and New York (1995).

[46] INTERNATIONAL ORGANIZATION FOR STANDARDIZATION, Quality Management Systems — Fundamentals and Vocabulary, ISO 9000:2015, ISO, Geneva (2015).

[47] Statute of the International Atomic Energy Agency, IAEA, Vienna (1990).

[48] Convention on Third Party Liability in the Field of Nuclear Energy of 29th July 1960, as amended by the Additional Protocol of 28th January 1964 and by the Protocol of 16th November 1982, OECD/NEA, Paris (2004).
See https://www.oecd-nea.org/jcms/pl_31788

[49] INTERNATIONAL NUCLEAR SAFETY ADVISORY GROUP, Probabilistic Safety Assessment, Safety Series No. 75-INSAG-6, IAEA, Vienna (1992).

[50] INTERNATIONAL COMMISSION ON RADIOLOGICAL PROTECTION, Basic Anatomical and Physiological Data for Use in Radiological Protection: Reference Values, Publication 89, Pergamon Press, Oxford and New York (2002).

[51] INTERNATIONAL COMMISSION ON RADIOLOGICAL PROTECTION, Reference Man: Anatomical, Physiological and Metabolic Characteristics, Publication 23, Pergamon Press, Oxford and New York (1976).

[52] INTERNATIONAL COMMISSION ON RADIOLOGICAL PROTECTION, Assessing Dose of the Representative Person for the Purpose of Radiation Protection of the Public and the Optimisation of Radiological Protection: Broadening the Process, ICRP Publication 101, Ann. ICRP **36** 3, Elsevier Science, Oxford (2006).

[53] INTERNATIONAL ATOMIC ENERGY AGENCY, Code of Conduct on the Safety of Research Reactors, IAEA/CODEOC/RR/2006, IAEA, Vienna (2006).

[54] INTERNATIONAL ATOMIC ENERGY AGENCY, Safety Classification of Structures, Systems and Components in Nuclear Power Plants, IAEA Safety Standards Series No. SSG-30, IAEA, Vienna (2014).

[55] INTERNATIONAL NUCLEAR SAFETY ADVISORY GROUP, Safety Culture, Safety Series No. 75-INSAG-4, IAEA, Vienna (1991).

[56] INTERNATIONAL ATOMIC ENERGY AGENCY, Dangerous Quantities of Radioactive Material (D-values), Emergency Preparedness and Response, EPR-D-VALUES 2006, IAEA, Vienna (2006).

[57] INTERNATIONAL ATOMIC ENERGY AGENCY, Predisposal Management of Radioactive Waste, IAEA Safety Standards Series No. GSR Part 5, IAEA, Vienna (2009).

[58] INTERNATIONAL ATOMIC ENERGY AGENCY, Volcanic Hazards in Site Evaluation for Nuclear Installations, IAEA Safety Standards Series No. SSG-21, IAEA, Vienna (2012).

[59] INTERNATIONAL ATOMIC ENERGY AGENCY, Classification of Radioactive Waste, IAEA Safety Standards Series No. GSG-1, IAEA, Vienna (2009).

BIBLIOGRAPHY

It is not intended, or indeed possible, that this glossary cover all terms that might be used in safety related publications. Many terms used in safety related publications originate in other specialized fields, such as computing, geology, meteorology and seismology. For most such technical terms, the reader is referred to specialized glossaries or dictionaries for the relevant fields. Some other safety related glossaries, dictionaries, etc., that may be of use are listed below.

AMERICAN NATIONAL STANDARDS INSTITUTE, Glossary of Terms in Nuclear Science and Technology, American Nuclear Society Standards Subcommittee on Nuclear Terminology Units ANS-9, American Nuclear Society, La Grange Park, IL (1986).

BORDERS' CONSULTING GROUP, Borders' Dictionary of Health Physics, www.hpinfo.org.

INTERNATIONAL ATOMIC ENERGY AGENCY (Vienna)

Safety Related Terms for Advanced Nuclear Plants, IAEA-TECDOC-626 (1991).

Terms for Describing New, Advanced Nuclear Power Plants, IAEA-TECDOC-936 (1997).

IAEA Safeguards Glossary (2022 Edition) (in preparation).

Radioactive Waste Management Glossary (2003), www-newmdb.iaea.org/

INTERNATIONAL COMMISSION ON RADIOLOGICAL PROTECTION (Pergamon Press, Oxford and New York)

Doses to the Embryo and Fetus from Intakes of Radionuclides by the Mother, Publication 88 (2001).

Basic Anatomical and Physiological Data for Use in Radiological Protection: Reference Values, Publication 89 (2002).

Guide for the Practical Application of the ICRP Human Respiratory Tract Model, Supporting Guidance 3, ICRP G3 (2003).

INTERNATIONAL ELECTROTECHNICAL COMMISSION, International Electrotechnical Vocabulary: Chapter 393 (Nuclear Instrumentation: Physical Phenomena and Basic Concepts), Rep. IEC 50(393), IEC, Geneva (1996).

INTERNATIONAL ORGANIZATION FOR STANDARDIZATION (Geneva)

Nuclear Energy: Vocabulary (Second Edition), ISO 921:1997 (1997).

ISO/IEC Guide 99:2007: International Vocabulary of Metrology: Basic and General Concepts and Associated Terms (VIM) (2007).

ANNEX:
SI UNITS AND PREFIXES

■	SI base units	(See International Standard
▪	SI derived units and non-SI units accepted for use with SI	ISO 1000 and the several
▫	Additional units accepted for use with SI for the time being	parts of ISO 31.)

Prefixes for SI (and metric units)

d	(deci)	10^{-1}	da	(deca)	10^{1}
c	(centi)	10^{-2}	h	(hecto)	10^{2}
m	(milli)	10^{-3}	k	(kilo)	10^{3}
μ	(micro)	10^{-6}	M	(mega)	10^{6}
n	(nano)	10^{-9}	G	(giga)	10^{9}
p	(pico)	10^{-12}	T	(tera)	10^{12}
f	(femto)	10^{-15}	P	(peta)	10^{15}
a	(atto)	10^{-18}	E	(exa)	10^{18}

Length
- ■ m metre
- ▫ Å ångström (10^{-10} m)

Area
- ▫ a are (10^2 m^2)
- ▫ ha hectare (10^4 m^2)
- ▫ b barn (10^{-28} m^2)

Volume
- ▪ L litre

Mass
- ■ kg kilogram
- ▪ t tonne (10^3 kg)
- ▪ u unified atomic mass unit
- ▪ T tesla

Time
- ■ s second
- ▪ min minute
- ▪ h hour
- ▪ d day

Temperature
- ■ K kelvin
- ▪ °C degree Celsius

Pressure
(Indicate absolute (abs) or gauge (g) as required, e.g. 304 kPa (g))

- ▪ Pa pascal (N/m^2)
- ▫ bar bar (10^5 Pa)

Radiation units
- ▪ Bq becquerel (dimensions: s^{-1})
- ▪ Gy gray (1 Gy = 1 J/kg)
- ▪ Sv sievert
- ▫ Ci curie (1 Ci = 37 GBq)
- ▫ R röntgen (1 R = 258 μC/kg)
- ▫ rad rad (100 rad = 1 Gy)
- ▫ rem rem (100 rem = 1 Sv)

Electricity and magnetism
- ■ A ampere
- ▪ C coulomb
- ▪ eV electronvolt
- ▪ F farad
- ▪ H henry
- ▪ Hz hertz (cycles per second)
- ▪ Ω ohm
- ▪ S siemens (ohm^{-1})
- ▪ V volt
- ▪ W watt
- ▪ Wb weber

Others
- ■ cd candela
- ■ mol mole
- ▪ J joule
- ▪ lm lumen
- ▪ lx lux
- ▪ N newton
- ▪ rad radian
- ▪ sr steradian
- ▪ ° degree of angle
- ▪ ′ minute of angle
- ▪ ″ second of angle

IAEA
International Atomic Energy Agency

ORDERING LOCALLY

IAEA priced publications may be purchased from the sources listed below or from major local booksellers.

Orders for unpriced publications should be made directly to the IAEA. The contact details are given at the end of this list.

NORTH AMERICA

Bernan / Rowman & Littlefield

15250 NBN Way, Blue Ridge Summit, PA 17214, USA
Telephone: +1 800 462 6420 • Fax: +1 800 338 4550

Email: orders@rowman.com • Web site: www.rowman.com/bernan

REST OF WORLD

Please contact your preferred local supplier, or our lead distributor:

Eurospan Group

Gray's Inn House
127 Clerkenwell Road
London EC1R 5DB
United Kingdom

Trade orders and enquiries:

Telephone: +44 (0)176 760 4972 • Fax: +44 (0)176 760 1640
Email: eurospan@turpin-distribution.com

Individual orders:

www.eurospanbookstore.com/iaea

For further information:

Telephone: +44 (0)207 240 0856 • Fax: +44 (0)207 379 0609
Email: info@eurospangroup.com • Web site: www.eurospangroup.com

Orders for both priced and unpriced publications may be addressed directly to:

Marketing and Sales Unit
International Atomic Energy Agency
Vienna International Centre, PO Box 100, 1400 Vienna, Austria
Telephone: +43 1 2600 22529 or 22530 • Fax: +43 1 26007 22529
Email: sales.publications@iaea.org • Web site: www.iaea.org/publications

22-04030E